The Essential Guide to Technical Product Specification:

Engineering Drawing

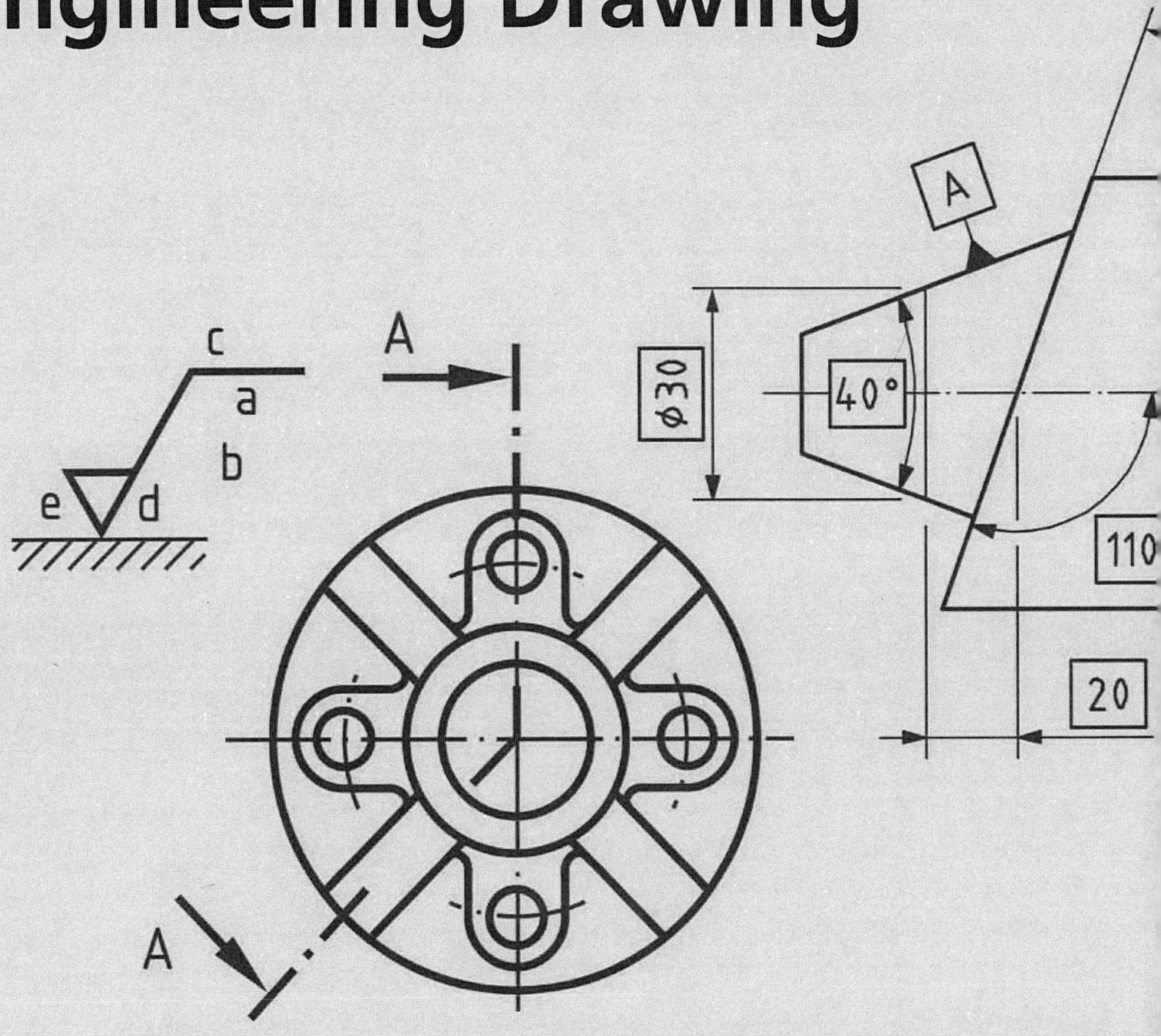

BSI

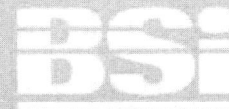

The Essential Guide to Technical Product Specification: Engineering Drawing

The Essential Guide to Technical Product Specification: Engineering Drawing

Colin Simmons
and
Neil Phelps

First published in the UK in 2009 by

BSI
389 Chiswick High Road
London W4 4AL

© British Standards Institution 2009

All rights reserved. Except as permitted under the *Copyright, Designs and Patents Act 1988*, no part of this publication may be reproduced, stored in a retrieval system or transmitted in any form or by any means – electronic, photocopying, recording or otherwise – without prior permission in writing from the publisher.

Whilst every care has been taken in developing and compiling this publication, BSI accepts no liability for any loss or damage caused, arising directly or indirectly in connection with reliance on its contents except to the extent that such liability may not be excluded in law.

While every effort has been made to trace all copyright holders, anyone claiming copyright should get in touch with the BSI at the above address.

BSI has no responsibility for the persistence or accuracy of URLs for external or third-party internet websites referred to in this book, and does not guarantee that any content on such websites is, or will remain, accurate or appropriate.

The right of Colin Simmons and Neil Phelps to be identified as the authors of this Work has been asserted by them in accordance with sections 77 and 78 of the *Copyright, Designs and Patents Act 1988*.

Typeset in Optima and Gill Sans by Monolith – http://www.monolith.uk.com
Printed in Great Britain by Berforts Group, Stevenage

British Library Cataloguing in Publication Data
A catalogue record for this book is available from the British Library

ISBN 978 0 580 62673 9

Contents

Introduction		**vii**
Dimensioning and tolerancing of size		**1**
1.1	Introduction	1
1.2	General principles	1
1.3	Types of dimension	2
1.4	Dimensioning conventions	3
1.5	Arrangement of dimensions	4
1.6	Methods for dimensioning common features	9
1.7	Dimensioning screw threads and threaded parts	12
1.8	Dimensioning chamfers and countersinks	13
1.9	Equally spaced repeated features	14
1.10	Dimensioning of curved profiles	16
1.11	Dimensioning of keyways	17
1.12	Tolerancing	18
1.13	Interpretations of limits of size for a feature-of-size	19
1.14	Datum surfaces and functional requirements	21
1.15	Relevant standards	21
Geometric tolerancing datums and datum systems		**23**
2.1	Introduction	23
2.2	Terms and definitions	23
2.3	Basic concepts	26
2.4	Symbols	27
2.5	Tolerance frame	29
2.6	Toleranced features	29
2.7	Tolerance zones	32
2.8	Datums and datum systems	37
2.9	Supplementary indications	45
2.10	Examples of geometrical tolerancing	64
2.11	Relevant standards	114
Graphical symbols for the indication of surface texture		**115**
3.1	Introduction	115
3.2	The basic graphical symbol	115
3.3	Expanded graphical symbols	115
3.4	Mandatory positions for the indication of surface texture requirements	116
3.5	Surface texture parameters	117
3.6	Indication of special surface texture characteristics	118
3.7	Indications on drawings	120
3.8	Relevant standards	123

Welding, brazed and soldered joints – Symbolic representation		**125**
4.1	Introduction	125
4.2	Relevant standards	133
Limits and fits		**135**
5.1	Introduction	135
5.2	Selected ISO fits – Hole basis	135
5.3	Selected ISO fits – Shaft basis	138
5.4	Methods of specifying required fits	140
5.5	Relevant standards	140
Metric screw threads		**141**
6.1	Introduction	141
6.2	Thread designation	141
6.3	Relevant standards	168
Illustrated index to BS 8888		**169**
	Normative references	169

Introduction

This guide has been produced as a companion to BS 8888, presenting up-to-date information based on the technical product specification aspects of BS 8888 and the essential standards it references.

Its aim is to offer straightforward guidance together with pictorial representations, to all practitioners of technical product specification, i.e. those currently using BS 8888 and those who, in a bid to conform to global ISO practices, are making or wish to make, the transition from the old BS 308 to BS 8888.

Its scope is to provide the necessary tools to enable engineers engaged in design specification, manufacturing and verification with the essential basic information required for specifying a product or component.

It includes comprehensive sections extracted from and referenced to international standards relating to linear, geometric and surface texture dimensioning and tolerancing, together with the practice of welding symbology, limits and fits and thread data. It also includes an illustrated index to all standards referenced in BS 8888.

This guide does not replace BS 8888 which is the definitive standard for technical product realization.

Any element of BS 8888 not included in this guide should not be considered as less important to technical specification than those included.

Most of the drawings in this guide have been extracted (and adapted) from the following BSI publications: BS EN ISO 1101, BS EN ISO 1302, BS ISO 5459, BS 8888 and PP 8888, Parts 1 and 2.

Chapter 1

Dimensioning and tolerancing of size

1.1 Introduction

Dimensioning is the process of applying measurements to a technical drawing. It is crucial to the whole process by which the designer will communicate the information required for the manufacture and verification of products.

1.2 General principles

Dimensions shall be applied to the drawing accurately, clearly and unambiguously. The following points shall be regarded as general dimensioning principles to be applied to all technical drawings.

- Each dimension necessary for the definition of the finished product shall be shown once only.
- Never calculate a dimension from the other dimensions shown on the drawing, nor scale the drawing.
- There shall be no more dimensions than are necessary to completely define the product.
- Preferred sizes shall be used whenever possible (see notes).
- Linear dimensions shall be expressed in millimetres (unit symbol 'mm'). If this information is stated on the drawing, the unit symbol 'mm' may be omitted. If other units are used, the symbols shall be shown with their respective values.
- Dimensions shall be expressed to the least number of significant figures, e.g. 45 not 45,0.
- The decimal marker shall be a bold comma, given a full letter space and placed on the baseline.
- Where four or more numerals are to the left or right of the decimal marker, a full space shall divide each group of three numerals, counting from the position of the decimal marker, e.g. 400 or 100 but 12 500 (see notes).
- A zero shall precede a decimal of less than one, e.g. 0,5.
- An angular dimension shall be expressed in degrees and minutes, e.g. 20° and 22° 30′ or, alternatively, as a decimal, e.g. 30,5°.
- A full space shall be left between the degree symbol and the minute numeral.
- When an angle is less than one degree, it shall be preceded by a zero, e.g. 0° 30′.

NOTES: Preferred sizes are those referring to standard material stock sizes and standard components such as nuts, bolts, studs and screws.

Decimal marker points or commas are not used to separate groups of numerals. This causes ambiguity since the decimal marker is denoted by a comma.

1.3 Types of dimension

For the purposes of this section, the following definitions apply.

dimension
numerical value expressed in appropriate units of measurement and indicated graphically on technical drawings with lines, symbols and notes
  Dimensions are classified according to the following types.

functional dimension
dimension that is essential to the function of the piece or space ('F' in Figure 1). See also 1.14

non-functional dimension
dimension that is not essential to the function of the piece or space ('NF' in Figure 1)

auxiliary dimension
Dimension, given for information purposes only, that does not govern production or inspection operations and is derived from other values shown on the drawing or in related documents

NOTE: An auxiliary dimension is given in parentheses and no tolerance may be applied to it ('AUX' in Figure 1).

feature
individual characteristic such as a flat surface, a cylindrical surface, two parallel surfaces, a shoulder, a screw thread, a slot or a profile

end product
complete part ready for assembly or service
or
configuration produced from a drawing specification
or
part ready for further processing (for example, a product from a foundry or forge) or a configuration needing further processing

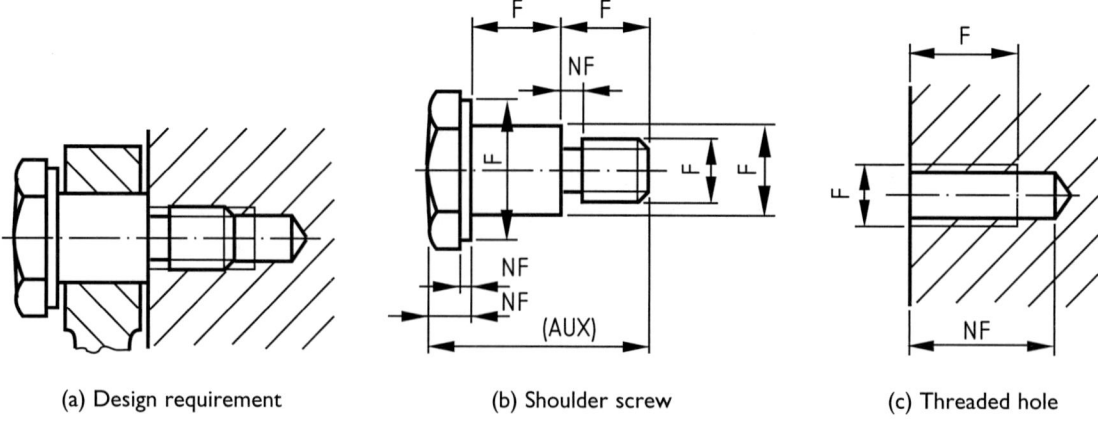

(a) Design requirement   (b) Shoulder screw   (c) Threaded hole

Figure 1 – Types of dimensioning

1.4 Dimensioning conventions

Technical product specification standards specify the following conventions when dimensioning drawings.

Extension lines shall normally be placed outside the view to aid clarity, as shown in Figure 2.

The extension line connects the dimension line (on which the value of the measurement is placed) to the reference points on the outline of the drawing. The following standard practice is specified.

Crossing of extension lines shall be avoided whenever possible.

There should be a small gap between the outline of the drawing and a projection line. The extension line shall extend slightly beyond the dimension line, as shown in Figure 2.

Extension lines shall, where possible, be drawn at right angles to the dimension line.

Centre-lines, extensions of centre-lines and continuations of outlines shall never be used as dimension lines. They may, however, be used as projection lines.

Arrowheads and origin circles are commonly used as terminators for dimension lines. Oblique strokes and points can also be used, as shown in Figures 3 and 4.

Dimension lines shall be unbroken even if the feature they refer to is shown as interrupted, as illustrated in Figure 5.

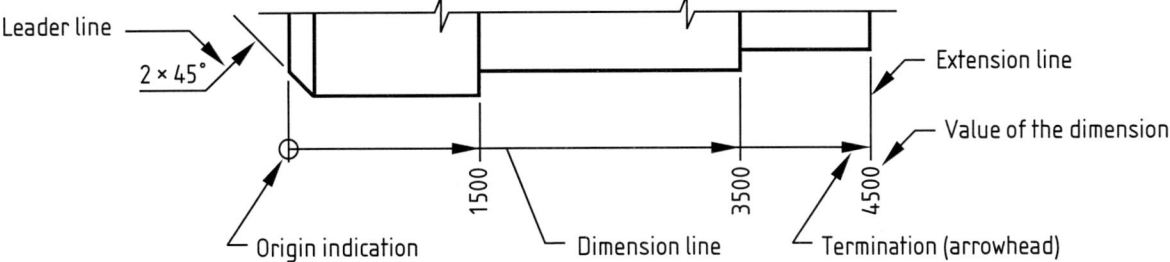

Figure 2 – Examples of extension lines and dimension lines

Terminators: dimension lines shall be terminated according to one of the representations shown in Figure 3.

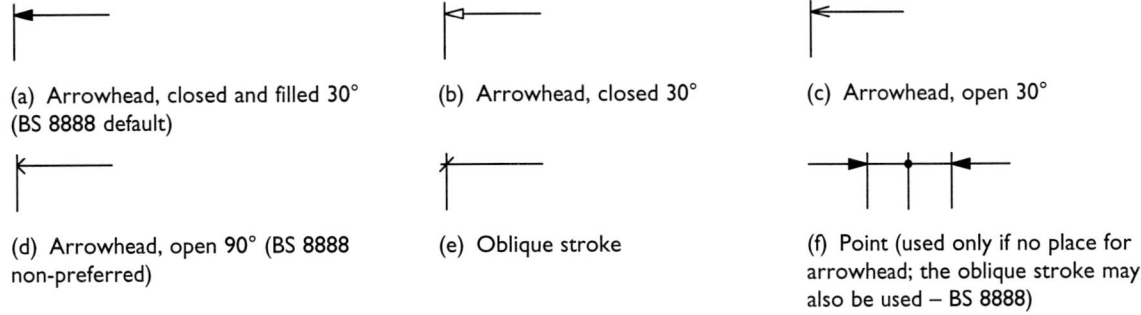

(a) Arrowhead, closed and filled 30° (BS 8888 default)

(b) Arrowhead, closed 30°

(c) Arrowhead, open 30°

(d) Arrowhead, open 90° (BS 8888 non-preferred)

(e) Oblique stroke

(f) Point (used only if no place for arrowhead; the oblique stroke may also be used – BS 8888)

Figure 3 – Terminators for dimension lines

Origin indication: the origin of the dimension line shall be indicated as shown in Figure 4.

Figure 4 – Origin indication

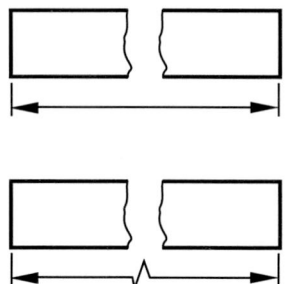

Figure 5 – Dimensioning interrupted features

When symmetrical parts are drawn partially, the portions of the dimension lines shall extend a short way beyond the axis of symmetry and the second termination shall be omitted, as shown in Figure 6.

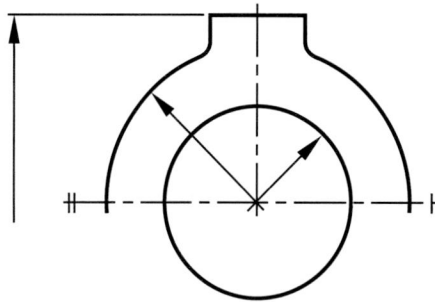

Figure 6 – Dimension lines on a partial view of a symmetrical part

1.5 Arrangement of dimensions

The way in which dimensions are typically used on drawings is shown in Figure 7. Conventions for arranging dimensions on drawings are as follows.

Dimensions shall be placed in the middle of the dimension line above and clear of it.

Dimensions shall not be crossed or separated by other lines on the drawing.

Values of angular dimensions shall be oriented so that they can be read from the bottom or the right-hand side of the drawing, as shown in Figure 8.

Where space is limited, the dimension can be placed centrally, above, or in line with, the extension of one of the dimension lines, as shown in Figure 9.

Larger dimensions shall be placed outside smaller dimensions, as shown in Figure 10.

Dimensions of diameters shall be placed on the view that provides the greatest clarity, as shown in Figure 11.

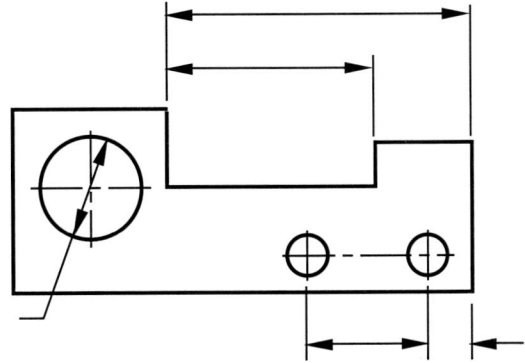

Figure 7 – Examples of the ways in which dimensions are typically used on drawings

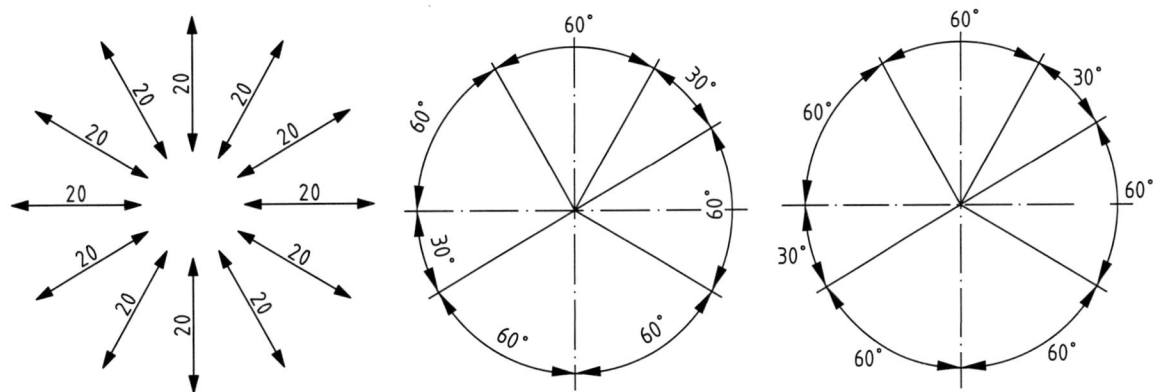

Figure 8 – Orientation of linear and angular dimensions

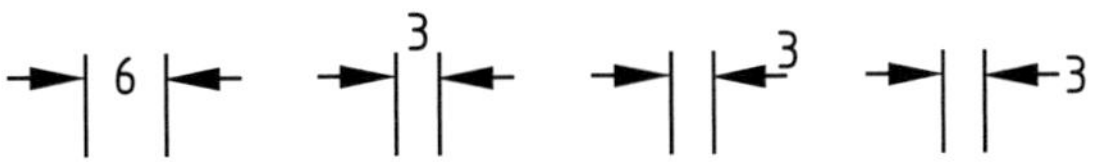

Figure 9 – Dimensioning smaller features

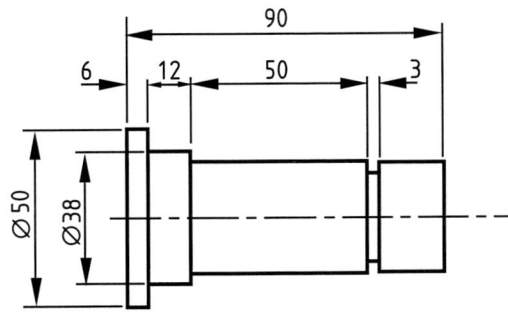

Figure 10 – Larger dimensions placed outside smaller dimensions

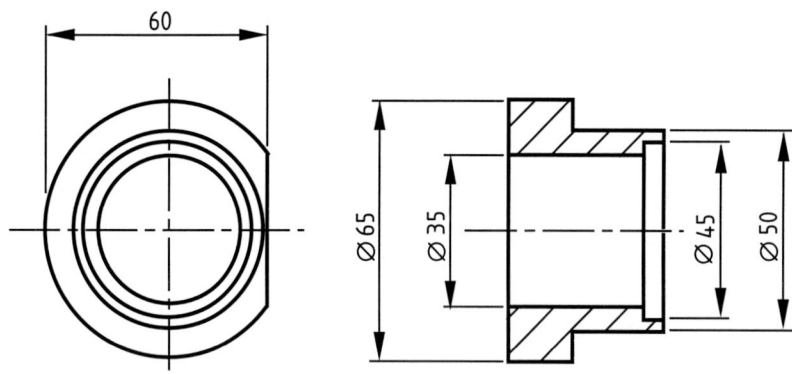

Figure 11 – Dimensions of diameters placed on view providing greatest clarity

Dimensioning from a common feature can be used where a number of dimensions of the same direction relate to a common origin.

Dimensioning from a common feature may be executed as parallel dimensioning or as superimposed running dimensioning.

Parallel dimensioning is the placement of a number of single dimension lines parallel to one another and spaced out so that the dimensional value can easily be added in, as shown in Figure 12a.

Superimposed running dimensioning is a simplified parallel dimensioning and may be used where there are space limitations. The common origin is as shown in Figure 12. Dimension values may be above and clear of the dimension line, as shown in Figure 12b; or in line with the corresponding extension line, as shown in Figure 12c.

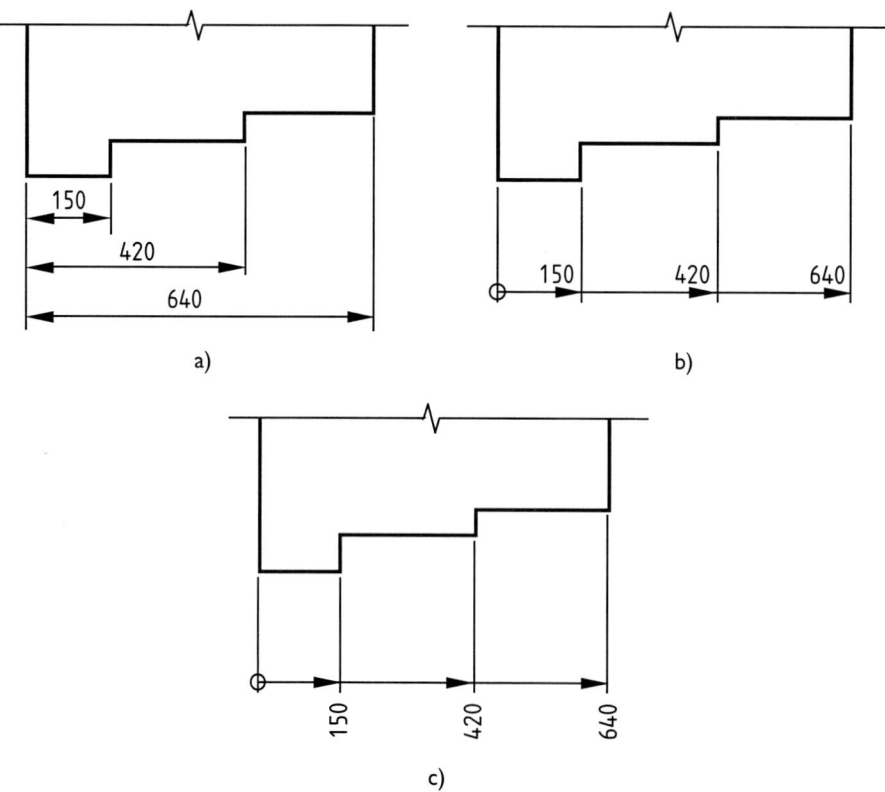

Figure 12 – Parallel dimensioning and running dimensioning

Dimensioning and tolerancing of size

Chain dimensioning consists of a chain of dimensions. These shall only be used where the possible accumulation of tolerances does not affect the function of the part, as shown in Figure 13.

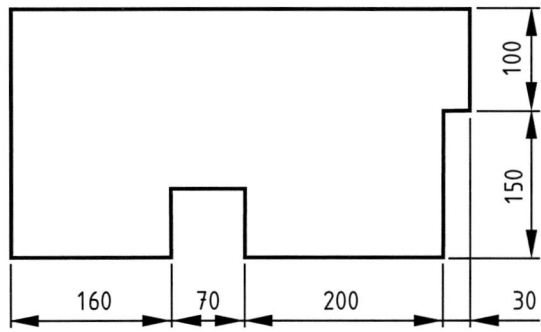

Figure 13 – Chain dimensioning

Combined dimensioning uses chain dimensioning and parallel dimensioning on the same drawing view. Figure 14a illustrates combining single dimensions and parallel dimensioning from a common feature. Figure 14b illustrates combining single dimensions and chain dimensions.

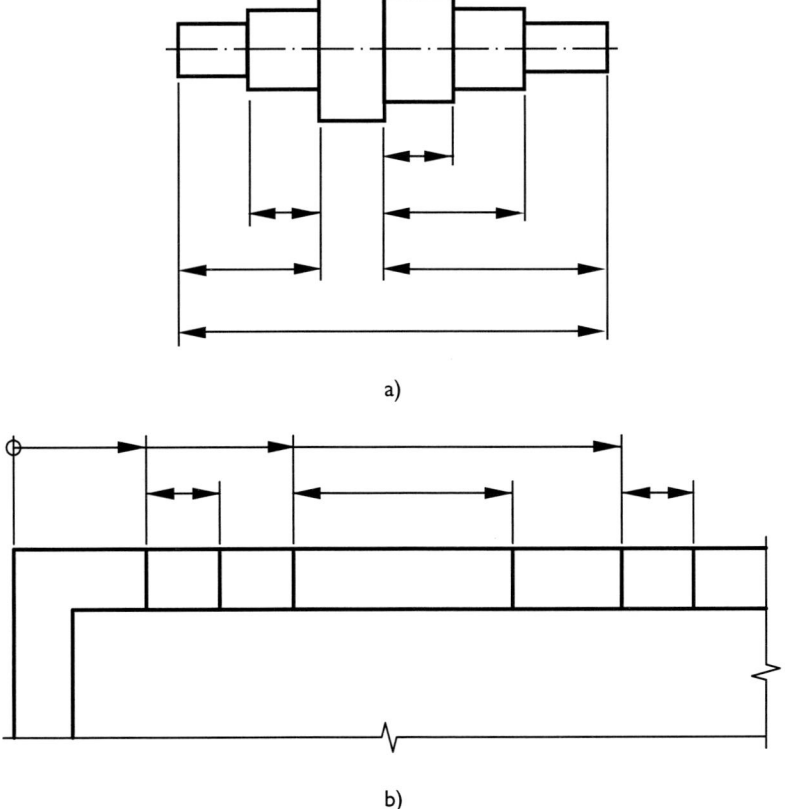

Figure 14 – Combined dimensioning

Dimensioning by coordinates uses superimposed running dimensioning in two directions at right angles, as shown in Figure 15a. The common origin may be any suitable common reference feature. It may be useful, instead of dimensioning as shown in Figure 15a, to tabulate dimensional values as shown in Figure 15b.

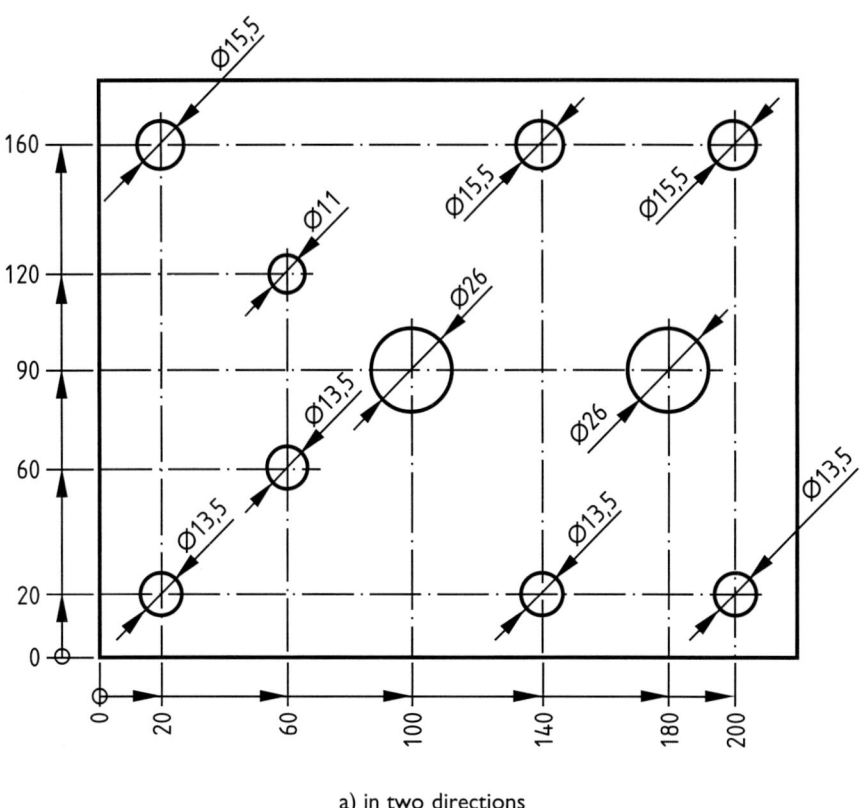

a) in two directions

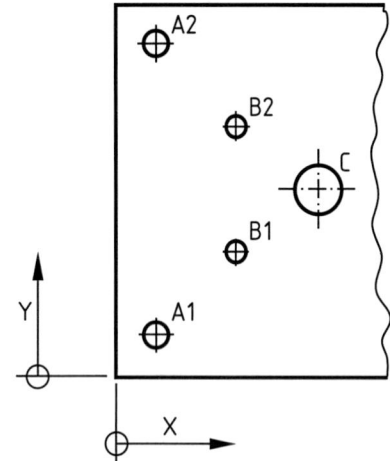

Hole	X	Y	Ø
A1	20	20	15
A2	20	160	15
B1	60	60	10
B2	60	120	10
C	100	90	25

b) tabulated

Figure 15 – Dimensioning by coordinates

1.6 Methods for dimensioning common features

Certain features, such as diameters, radii, squares, hole sizes, chamfers, countersinks and counter-bores, can occur frequently in engineering drawings.

A diameter of a circle or cylinder shall be dimensioned by prefixing the value with the symbol Ø, as shown in Figure 16. A square feature shall be dimensioned by prefixing the value with the symbol □. Additionally, square and flat features can be indicated by continuous narrow lines drawn diagonally on the flat feature, as shown in Figure 18.

Where dimension lines and other lines (e.g. extension lines) would otherwise intersect, the dimension lines to the feature can be dimensioned by leader lines as shown in Figure 16.

Where the whole view is not shown, concentric diameters shall be dimensioned as in Figure 17.

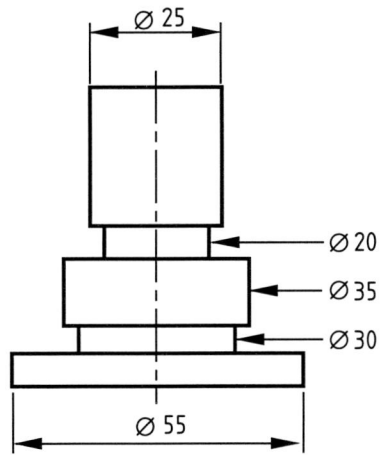

Figure 16 – Diameter dimensions indicated by leader lines

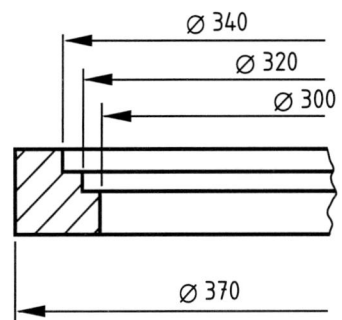

Figure 17 – Dimensioning concentric diameters on a partial view

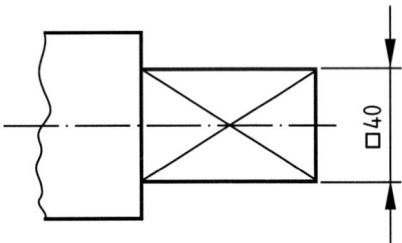

Figure 18 – Dimensioning a square

Circles shall be dimensioned as shown in Figure 19 and spherical surfaces as shown in Figure 20.

Radii of features shall be dimensioned by prefixing the value with the letter R. Radii shall be dimensioned by a line that passes through, or is in line with, the centre of the arc. The dimension line shall have one arrowhead only, which shall touch the arc.

Radii that require their centres to be located shall be dimensioned as in Figure 21a; those that do not shall be dimensioned as in Figure 21b. Spherical radii shall be dimensioned as shown in Figures 21c and 21d.

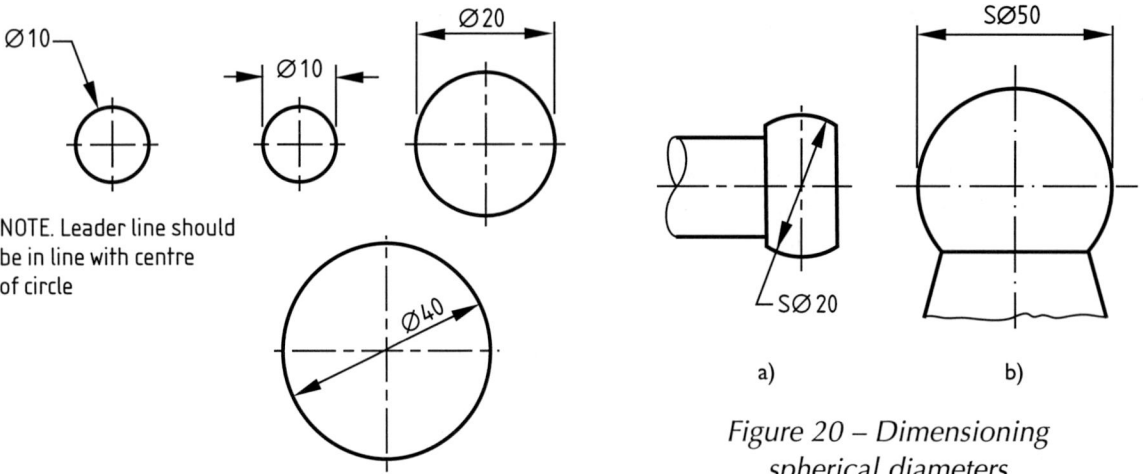

Figure 19 – Dimensioning a diameter

Figure 20 – Dimensioning spherical diameters

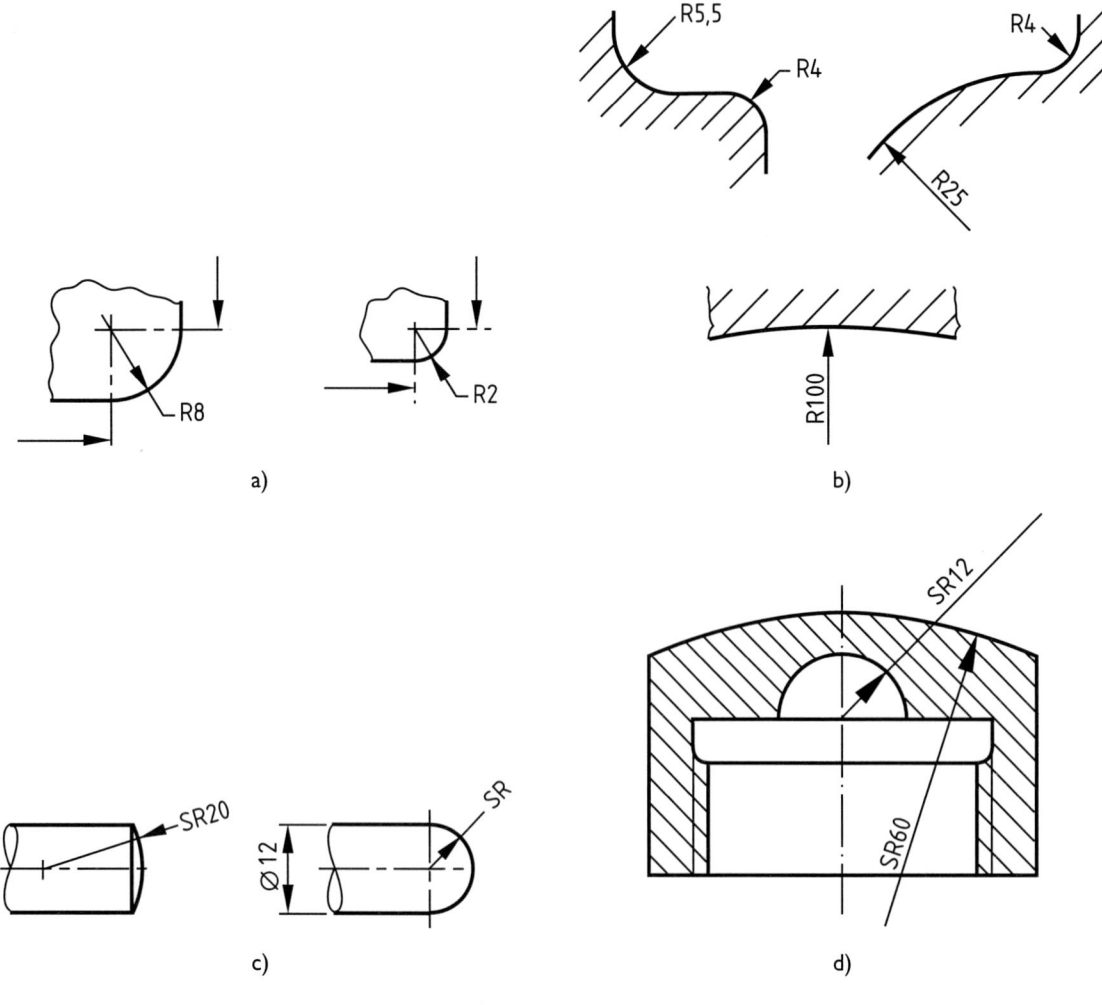

Figure 21 – Dimensioning radii

Holes shall be dimensioned as shown in Figure 22. The depth of the drilled hole, when given after the diameter, refers to the depth of the cylindrical portion of the hole and not to the extremity made by the point of the drill, unless otherwise specified.

The method of production (e.g. drill, punch, bore or ream) shall not be specified except where it is essential to the function of the part.

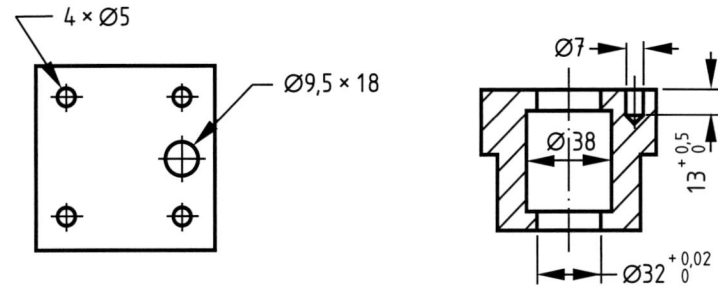

Figure 22 – Dimensioning holes

The dimensioning of chords, arcs and angles shall be as shown in Figure 23.

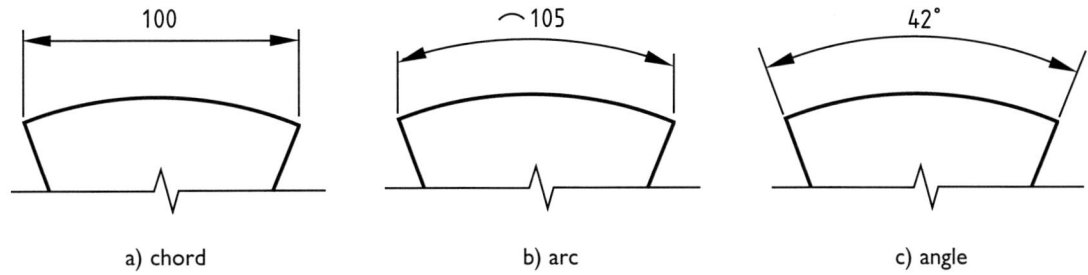

Figure 23 – Dimensioning chords, arcs and angles

Dimensioning the spacing of holes and other features on a curved surface shall be as shown in Figure 24, whether the dimensions are chordal or circumferential, they shall be indicated clearly on the drawing.

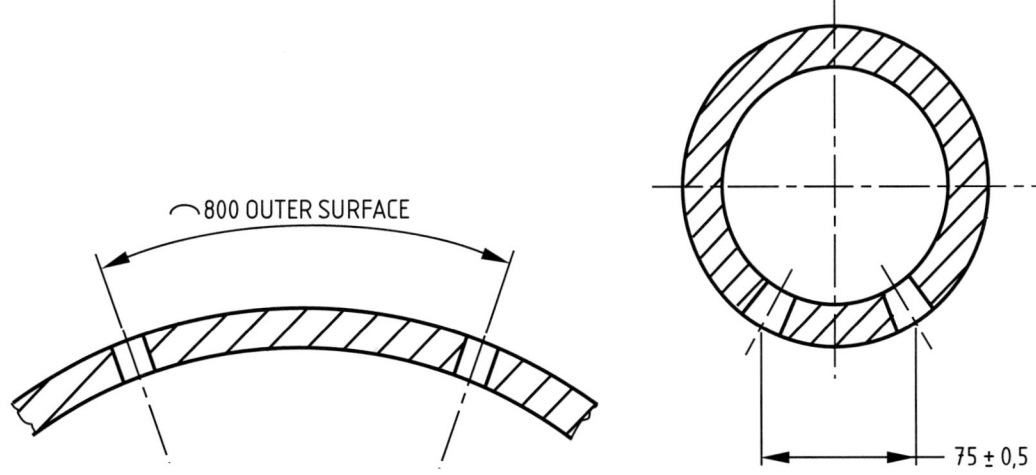

Figure 24 – Dimensions on a curved surface

1.7 Dimensioning screw threads and threaded parts

ISO metric screw threads shall be designated in accordance with BS EN ISO 6410-1, which specifies that the designation shall indicate the thread system, nominal diameter and the thread tolerance class. If necessary, the pitch shall also be indicated; however, when designating metric coarse threads, the pitch is generally omitted.

The nominal diameter refers to the major diameter of external and internal threads; the dimension relating to the depth of thread refers to the full depth of thread. The direction of a right hand thread (RH) is not generally noted; however left hand threads shall be denoted with the abbreviation 'LH' after the thread designation.

Thread system and size

The letter M, denoting ISO metric screw threads, shall be followed by the values of the nominal diameter and pitch (if required), with a multiplication sign between them, e.g. M8 × 1.

Thread tolerance class

For general use, the tolerance class 6H is suitable for internal threads and tolerance class 6g for external threads. The thread tolerance class shall be preceded by a hyphen, e.g. M10-6H or M10 × 1-6g.

Screw threads shall be dimensioned as shown in Figures 25 and 26.

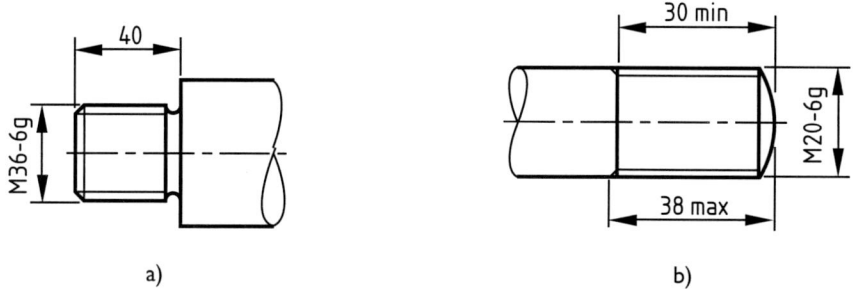

Figure 25 – Dimensioning external screw threads

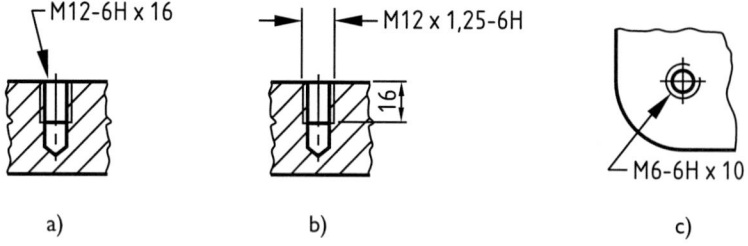

Figure 26 – Dimensioning internal screw threads

1.8 Dimensioning chamfers and countersinks

Chamfers shall be dimensioned as shown in Figure 27. Where the chamfer angle is 45°, the indications may be simplified as shown in Figure 28.

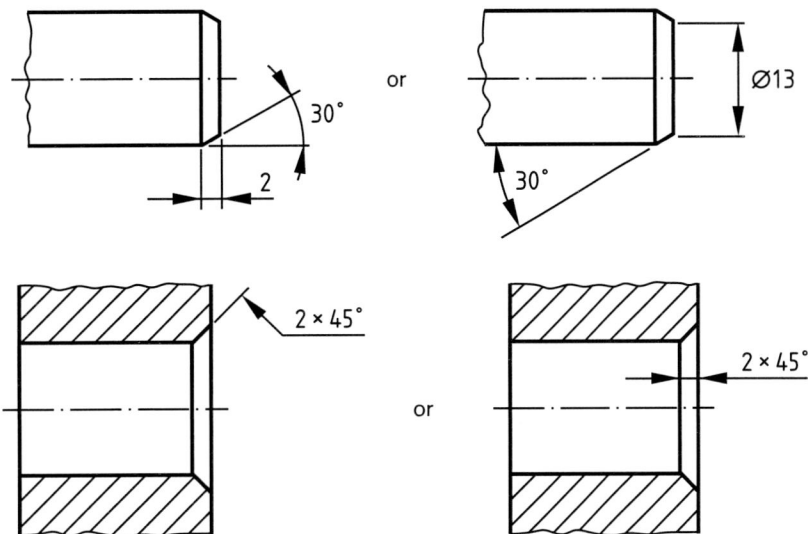

Figure 27 – Dimensioning external and internal chamfers

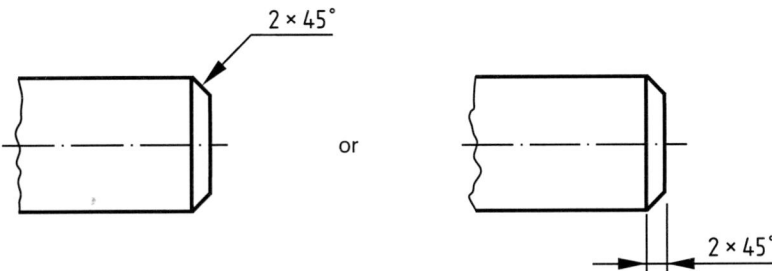

Figure 28 – Simplified dimensioning of chamfers

Countersinks shall be dimensioned by showing either the required diametral dimension at the included angle, or the depth and the included angle, as shown in Figure 29.

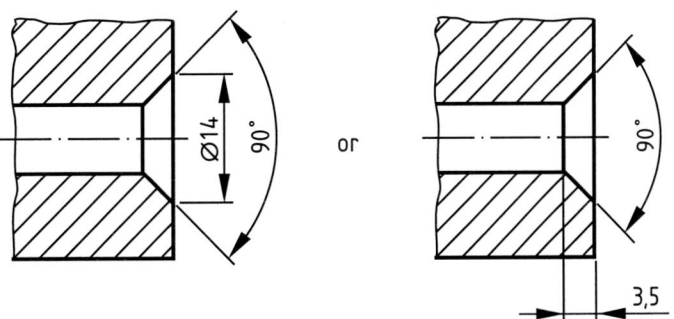

Figure 29 – Dimensioning countersinks

1.9 Equally spaced repeated features

Where repeated features are linearly spaced, a simplified method of dimensioning may be used, as shown in Figure 30.

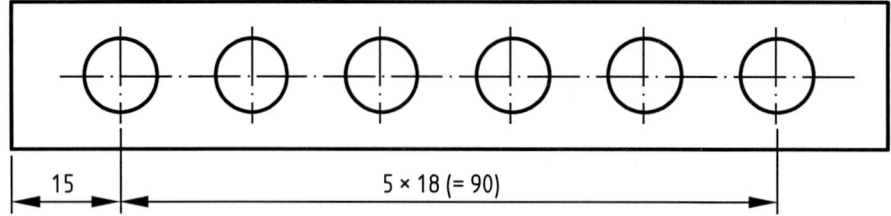

Figure 30 – Dimensioning of linear spacings

If there is any ambiguity, one feature space may be dimensioned as illustrated in Figure 31.

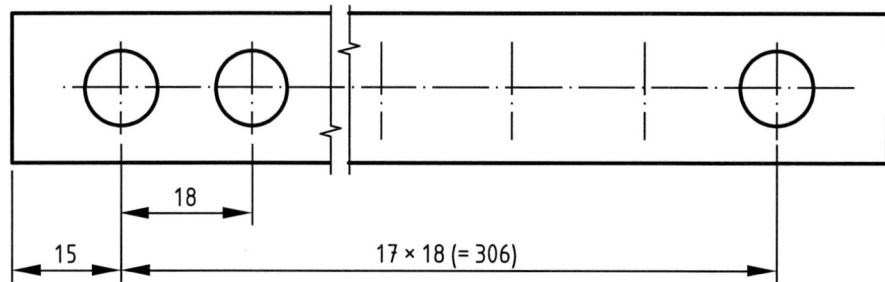

Figure 31 – Dimensioning of linear spacings to avoid confusion

Angular, equally spaced features shall be dimensioned as shown in Figure 32. The angle of the spacings can be omitted where the intent is explicit, as shown in Figure 33.

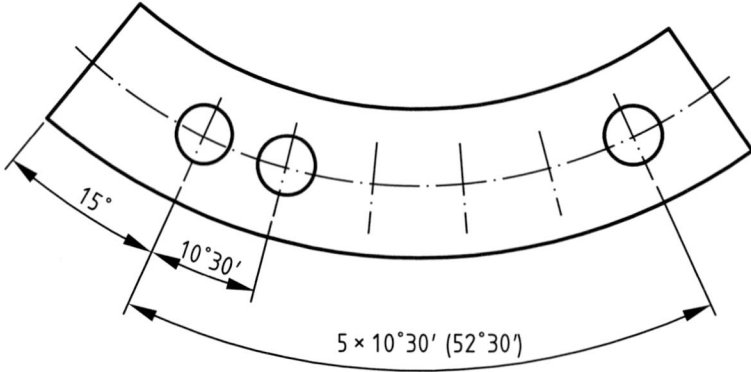

Figure 32 – Dimensioning angular spacing

Dimensioning and tolerancing of size

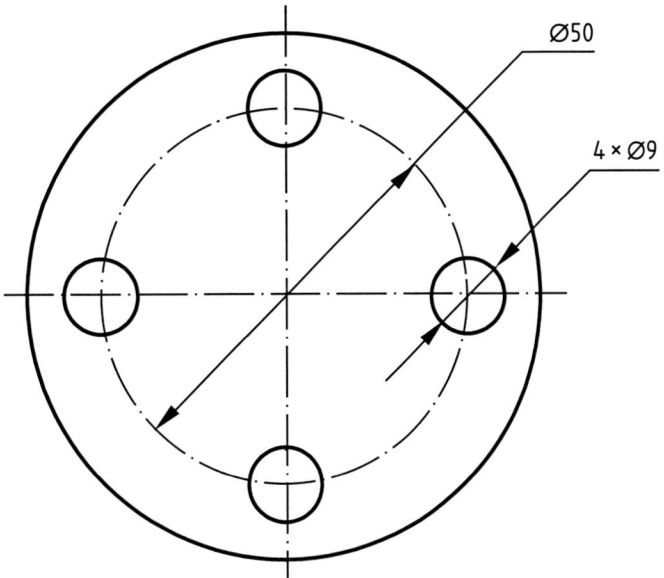

Figure 33 – Omission of angle of spacing

Circular spaced features can be dimensioned indirectly by specifying the number of common features as shown in Figure 34.

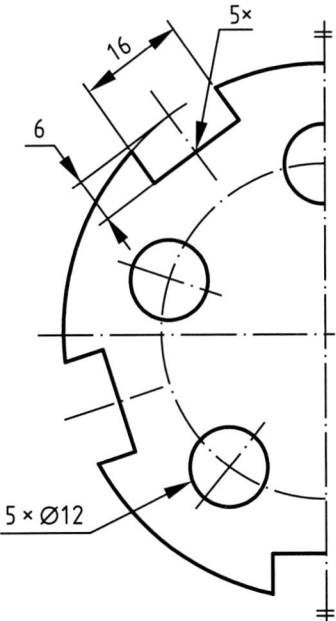

Figure 34 – Dimensioning circular spacings

Series or patterned features of the same size may be dimensioned as illustrated in Figures 35 and 36.

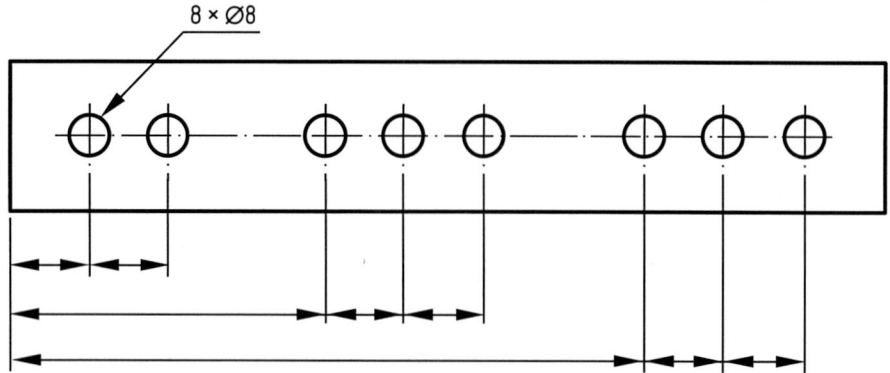

Figure 35 – Dimensioning a quantity of features of the same size – linear

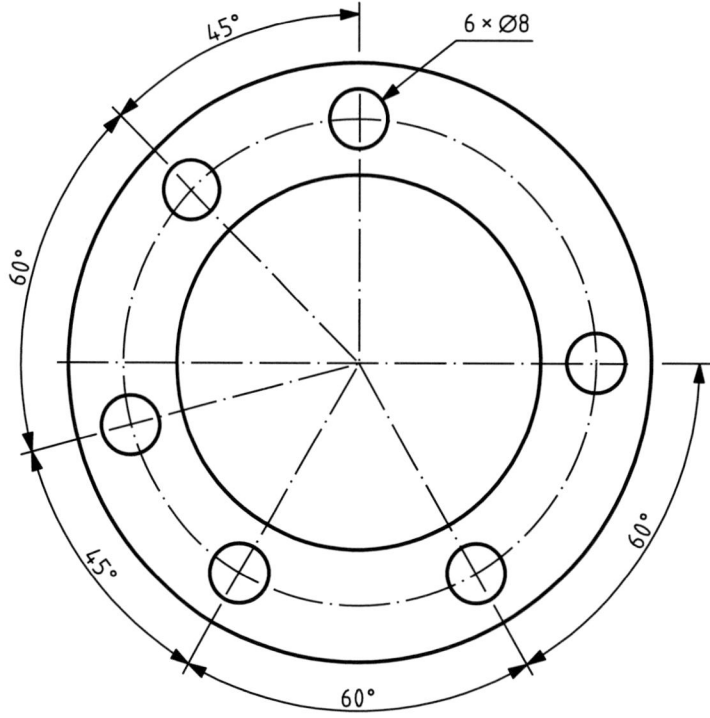

Figure 36 – Dimensioning a quantity of features of the same size – circular

1.10 Dimensioning of curved profiles

Curved profiles composed of circular arcs shall be dimensioned by radii, as shown in Figure 37.

Coordinates locating points on a curved surface, as shown in Figure 38, shall only be used when the profile is not composed of circular arcs. The more coordinates specified, the better the uniformity of the curve.

Dimensioning and tolerancing of size

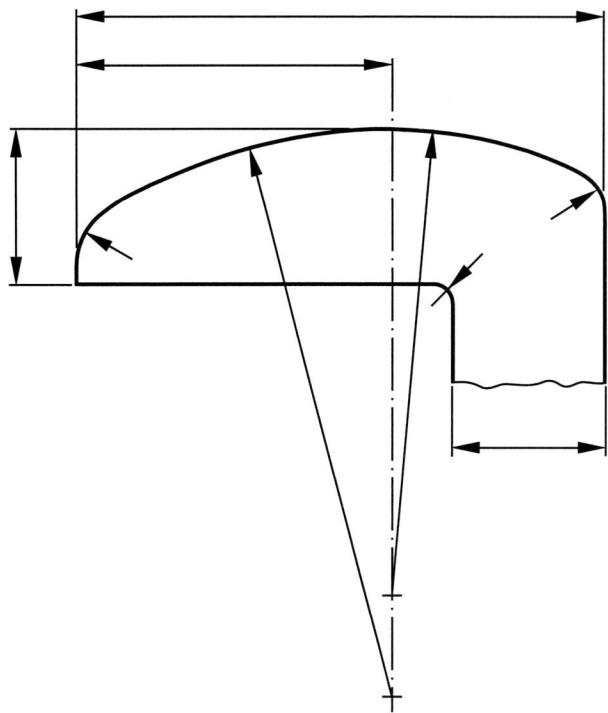

Figure 37 – Dimensioning of a curved profile

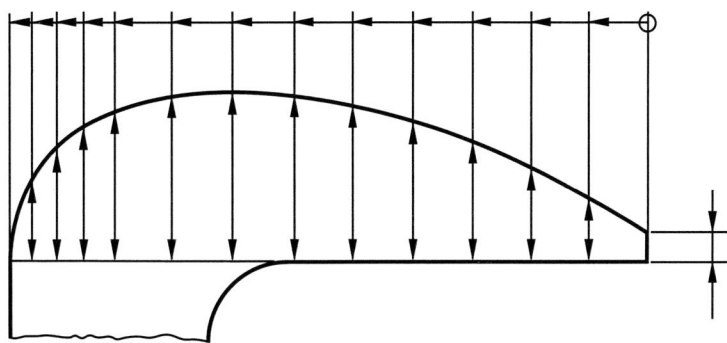

Figure 38 – Linear coordinates of a series of points through which a profile passes

1.11 Dimensioning of keyways

Keyways in hubs or shafts shall be dimensioned by one of the methods shown in Figure 39.

NOTE: Further information on keys and keyways is given in BS 4235-1, *Specification for metric keys and keyways – Part 1: Parallel and taper keys* and BS 4235-2, *Specification for metric keys and keyways – Part 2: Woodruff keys and keyways*.

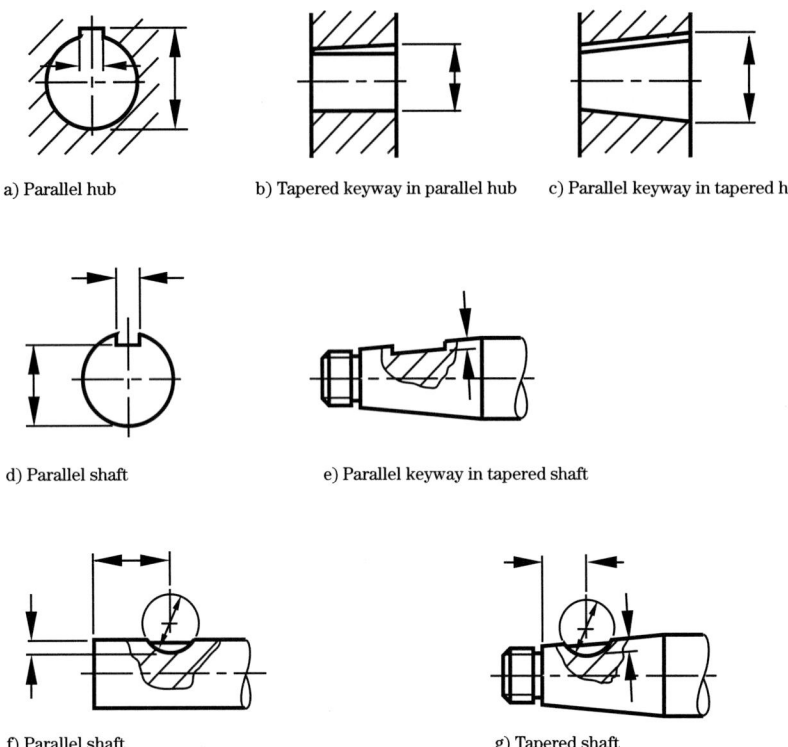

a) Parallel hub   b) Tapered keyway in parallel hub   c) Parallel keyway in tapered hub

d) Parallel shaft   e) Parallel keyway in tapered shaft

f) Parallel shaft   g) Tapered shaft

Figure 39 – Dimensioning of keyways

1.12 Tolerancing

Tolerancing is the practice of specifying the upper and lower limit for any permissible variation in the finished manufactured size of a feature. The difference between these limits is known as the tolerance for that dimension.

All dimensions (except auxiliary dimensions) are subject to tolerances.

Tolerances shall be specified for all dimensions that affect the functioning or interchange ability of the part.

Tolerances shall also be used to indicate where unusually wide variations are permissible.

Tolerances shall be applied either to individual dimensions or by a general note giving uniform or graded tolerances to classes of dimensions, for example:

TOLERANCE UNLESS OTHERWISE STATED   LINEAR ±0,4 ANGULAR ±0° 30′

The method shown in Figure 40a should be followed where it is required to tolerance individual linear dimensions. This method directly specifies both the limits of the size of the dimension, the tolerance being the difference between the limits of the size.

The larger limit of the size shall be placed above the smaller limit and both shall be given to the same number of decimal places.

The method shown in Figure 40b can be used as an alternative way of specifying tolerances.

Dimensioning and tolerancing of size

Figure 40 – Linear dimension tolerance by directly specifying limits of size

The methods shown in Figure 41 may be used to tolerance individual angular dimensions.

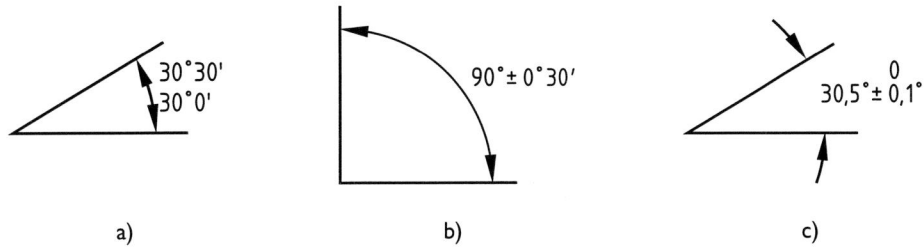

Figure 41 – Tolerancing angular dimensions

1.13 Interpretations of limits of size for a feature-of-size

Limits of size for an individual feature-of-size shall be interpreted according to the principles and rules defined in BS ISO 8015, BS EN ISO 14660-1 and BS EN ISO 14660-2.

A feature-of-size may consist of two parallel plane surfaces, a cylindrical surface or a spherical surface, in each case defined with a linear size. A feature-of-size may also consist of two plane surfaces at an angle to each other (a wedge) or a conical surface, in each case defined with an angular size.

BS ISO 8015 states that limits of size control only the actual local sizes (two-point measurements) of a feature-of-size and not its deviations of form (e.g. the roundness and straightness deviations of a cylindrical feature, or the flatness deviations of two parallel plane surfaces). Form deviations may be controlled by individually specified geometrical tolerances, general geometrical tolerances or through the use of the envelope requirement (where the maximum material limit of size defines an envelope of perfect form for the relevant surfaces; see BS ISO 8015).

BS ISO 8015 defines the principle of independency, according to which each specified dimensional and geometrical requirement on a drawing is met independently, unless a particular relationship is specified. A relationship may be specified through the use of the envelope requirement or material condition modifiers maximum material condition (MMC) or least material condition (LMC).

Where no relationship is specified, any geometrical tolerance applied to the feature-of-size applies regardless of feature size, and the two requirements shall be treated as unrelated, as shown in Figure 42. The limits of size do not control the form, orientation, or the spatial relationship between, individual features-of-size.

Consequently, if a particular relationship of size and form, or size and location, or size and orientation is required, it needs to be specified.

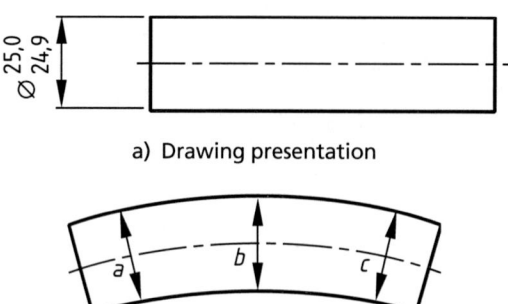

a) Drawing presentation

NOTE There is no form control (i.e. over roundness, straightness or cylindricity). Measurements a, b and c may lie between 25.0 mm and 24.9 mm, meeting the drawing requirement using two-point measurement only.

b) Permissible interpretation: straightness unconstrained

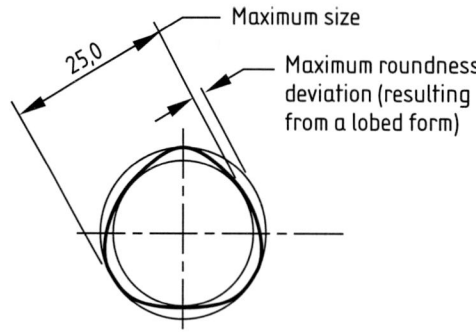

NOTE For any cross-section of the cylinder, there is no roundness control.

c) Permissible interpretation: roundness unconstrained

Figure 42 – Permissible interpretations when no form control is given on the drawing

1.13.1 Limits of size with mutual dependency of size and form

Some national standards apply, or have applied, the envelope requirement to all features-of-size by default. As the envelope requirement has been the default, they have not used a symbol to indicate this requirement; rather they use a note to indicate when this is not required. This system of tolerancing is sometimes described as the principle of dependency, or the application of the Taylor principle.

Standards which apply, or have applied, the envelope requirement by default include:

ASME Y14.5
The requirement that there shall be an envelope of perfect form corresponding to the maximum material size of the feature is defined as Rule #1).

BS 308
The principle of dependency was taken as the default option under BS 308, although the option of working to the principle of independency was included, through the use of the BS 308 triangle I indication.

Dimensioning and tolerancing of size

BS 8888

Prior to the 2004 revision; the principle of dependency was taken as the default option under BS 8888:2000 and BS 8888:2002, although the option of working to the principle of independency was included, through the use of the BS 8888 triangle I indication.

BS 8888:2004 and BS 8888:2006

the principle of dependency could be explicitly invoked through the use of the BS 8888 triangle D indication.

As the interaction between the envelope requirement and individual geometrical tolerances is not always fully defined within the ISO system, and as the application of the envelope requirement by default to all features-of-size is not formally supported within the ISO system, the use of the principle of dependency is no longer recommended.

1.14 Datum surfaces and functional requirements

Functional dimensions shall be expressed directly on the drawing, as shown in Figure 1. The application of this principle will result in the selection of reference features on the basis of the function of the product and the method of locating it in any assembly of which it may form a part.

If any reference feature other than one based on the function of the product is used, finer tolerances will be necessary to meet the functional requirement, which in turn will increase the cost of producing the product, as shown in Figure 43 on page 22.

1.15 Relevant standards

BS EN ISO 1660, *Technical drawings — Dimensioning and tolerancing of profiles*
BS ISO 129-1, *Technical drawings — Indication of dimensions and tolerances — Part 1: General principles*
BS ISO 3040, *Technical drawings — Dimensioning and tolerancing — Cones*
BS ISO 10579, *Technical drawings — Dimensioning and tolerancing — Non-rigid parts*
BS ISO 406, *Technical drawings — Tolerancing of linear and angular dimensions*
BS EN 22768-1, *General tolerances — Part 1: Tolerances for linear and angular dimensions without individual tolerance indications*
BS 4235-1, *Specification for metric keys and keyways — Part 1: Parallel and taper keys*
BS 4235-2, *Specification for metric keys and keyways — Part 2: Woodruff keys and keyways*
BS ISO 8015, *Technical drawings — Fundamental tolerancing principle*
BS EN ISO 14660-1, *Geometrical Product Specifications (GPS) — Geometrical features — Part 1: General terms and definitions*
BS EN ISO 14660-2, *Geometrical Product Specifications (GPS) — Geometrical features — Part 2: Extracted median line of a cylinder and a cone, extracted median surface, local size of an extracted feature*
PP 8888-2, *Engineering drawing practice: a guide for further and higher education to BS 8888:2006, Technical product specification (TPS)*

Description	Drawing
a) Assembly drawing showing a given functional requirement, namely the limits of height of the top face of item 1 above the top face of item 3, with a tolerance of 0.08 mm	
b) Detail of head of item 1 showing given limits of size, with a tolerance of 0.03 mm	
c) Item 2 dimensioned from a functional reference surface NOTE: One direct dimension with a tolerance of 0.05 mm is needed to satisfy the condition shown in a). A nominal flange thickness of 5 mm has been assumed. This value is non-functional and can have any large tolerance.	
d) Item 2 dimensioned from a non-functional reference surface NOTE: Tolerances have had to be reduced; two dimensions with tolerances of, say, 0.02 mm for the flange and 0.03 mm are now needed to satisfy the condition shown in a).	

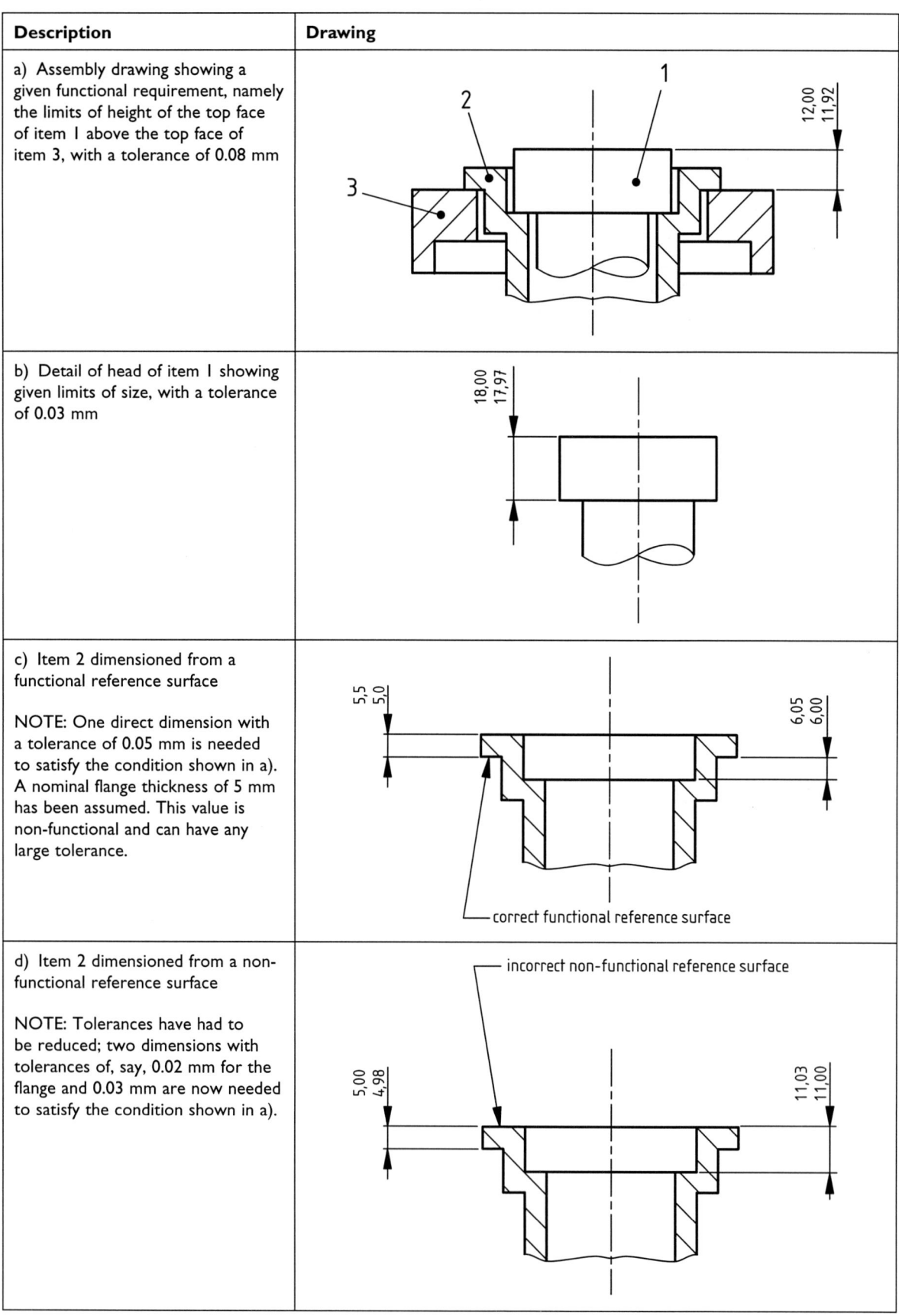

Figure 43 – Effect on tolerances by changing datum surfaces from those determined by functional requirements

Chapter 2

Geometric tolerancing datums and datum systems

2.1 Introduction

This section comprises information extracted from BS EN ISO 1101 and BS ISO 5459. Information not extracted should not be considered as being less important and for a more in-depth awareness of the application of geometric tolerancing and datums, both standards should be consulted.

Geometric tolerances have been developed to replace the 'written word' when specifying conditions which apply to a feature on a product. Geometric tolerances control the deviation of the feature from its theoretically exact form, orientation or location, regardless of the feature size.

Datums and datum systems are used as the basis for establishing the geometric relationship of related features.

2.2 Terms and definitions

annotation plane
conceptual plane containing annotation

NOTES: It is desirable that annotation planes intersect or be coincident with a model feature.

The plane is 'conceptual' because it is not physically shown as geometry on the model but is provided to replace the drawing media. [BS ISO 16792:2006, 3.2.2].

datum feature
non-ideal integral feature used for establishing a datum

NOTE: A datum feature can be a complete surface, a portion of a complete surface, or a feature of size.

associated feature (for establishing a datum)
ideal feature which is fitted to the datum feature with a specific association criterion

NOTES: The type of the associated feature is generally the same as the type of the nominal integral feature used to establish the datum.

The associated feature for establishing a datum simulates the contact between the real surface of the workpiece and other components of an assembly or the fixture used to locate or orientate the workpiece during manufacture or inspection.

collection surface

two or more surfaces considered simultaneously as a single surface

EXAMPLE 1: Two intersecting planes may be considered together or separately. When the two intersecting planes are considered simultaneously as a single surface, that surface is a collection surface.

collection plane

plane, established from a nominal feature on the workpiece, defining a closed compound continuous feature

NOTES: The collection plane may be required when the 'all around' symbol is applied.

The tolerance requirement applies to each surface or line element independently, unless otherwise specified (e.g. by using CZ symbol).

compound continuous feature

feature composed of several single features joined together without gaps

NOTES: A compound continuous feature can be closed or not closed.

A non closed compound continuous feature can be defined by the way of using the 'between' symbol.

A closed compound continuous feature can be defined by the way of using the 'all around' symbol. In this case, it is a set of single features whose intersection with any plane parallel to a collection plane is a line or a point.

single datum

datum established from one datum feature taken on a single surface or from one feature of size

NOTE: A single surface can be complex, prismatic, helical, cylindrical, revolute, planar or spherical.

common datum

datum established from two or more datum features considered simultaneously

NOTES: To define a common datum, it is necessary to consider the collection surface created by the considered datum features.

The collection surface can be complex, prismatic, helical, cylindrical, revolute, planar or spherical.

datum system

datum established from two or more datum features considered in a specific order

NOTES: To define a datum system, it is necessary to consider the collection surface created by the considered datum features.

The collection surface can be complex, prismatic, helical, cylindrical, revolute, planar or spherical.

datum target

portion of a datum feature which can be a point, a line or an area

derived feature

centre point, median line or median surface from one or more integral features

EXAMPLE 2: The centre of a sphere is a derived feature obtained from the sphere, which is an integral feature.

EXAMPLE 3: The median line of a cylinder is a derived feature obtained from the cylindrical surface, which is an integral feature.

direction feature

feature, established from an extracted feature of the workpiece, identifying the direction in which the tolerance value applies

NOTES: The direction feature can be a plane, a cylinder or a cone.

For a line in a surface the use of a direction feature makes it possible to change the direction of the width of the tolerance zone.

The direction feature is used on a complex surface or complex profile when the direction of the tolerance value is not normal to the specified geometry.

By default, the direction feature is a cone, a cylinder or a plane constructed from the datum or datum system defined in the tolerance frame. The geometry of the direction feature depends on the geometry of the toleranced feature.

extracted feature (integral)

approximated representation of the real (integral) feature, obtained by extracting a finite number of points from the real (integral) feature

extracted feature (derived)

centre point, median line or median surface derived from one or more extracted integral features

feature of size

geometrical shape defined by a linear or angular dimension which is a size

NOTE: The features of size can be a cylinder, a sphere, two parallel opposite surfaces, a cone or a wedge.
[BS EN ISO 14660-1:2000, 2.2].

geometrical feature

point, line or surface

integral feature

surface or line on a surface

intersection plane

plane, established from an extracted feature of the workpiece, identifying a line on an extracted surface (integral or median) or a point on an extracted line

NOTE: The use of intersection planes makes it possible to define toleranced features independent of the view.

orientation feature

feature, established from an extracted feature of the workpiece, identifying the orientation of the tolerance zone

NOTES: For a derived feature the use of an orientation feature makes it possible to define the direction of the width of the tolerance zone independently of the TED model (case of location) or of the datum (case of orientation).

The orientation feature is only used when the toleranced feature is a median feature (centre point, median straight line) and the tolerance zone is defined by two parallel straight lines or two parallel planes.

orientation plane

plane, established from an extracted feature of the workpiece, identifying the orientation of the width of the tolerance zone

NOTES: For a derived feature, the use of an orientation plane makes it possible to define the direction of the width of the tolerance zone independent of the view.

For a line in a surface, the use of an orientation plane makes it possible to change the direction of the width of the tolerance zone.

theoretically exact dimension

dimension indicated on technical product documentation, which is not affected by individual or general tolerances

NOTES: For the purpose of this guide, the term 'theoretically exact dimension' has been abbreviated to TED.

A TED can be a linear, circular or spherical dimension or an angular dimension. It is possible for example to write a theoretically exact angle.

A TED can define:

- the extension or the relative location of a portion of one feature or
- the length of the projection of a feature or
- the theoretical orientation or location from one or more features, or
- the nominal shape of a feature.

A TED is indicated by a dimension frame.

tolerance zone

space limited by one or several geometrically perfect lines or surfaces, and characterized by a linear dimension, called a tolerance

2.3 Basic concepts

Geometrical tolerances shall be specified in accordance with functional requirements. Manufacturing and inspection requirements can also influence geometrical tolerancing.

NOTE: Indicating geometrical tolerances on a drawing does not necessarily imply the use of any particular method of production, measurement or gauging.

A geometrical tolerance applied to a feature defines the tolerance zone within which that feature shall be contained.

A feature is a specific portion of the workpiece, such as a point, a line or a surface; these features can be integral features (e.g. the external surface of a cylinder) or derived (e.g. a median line or median surface). See BS EN ISO 14660-1.

According to the characteristic to be toleranced and the manner in which it is dimensioned, the tolerance zone is one of the following:

- the space within a circle;
- the space between two concentric circles;
- the space between two equidistant lines or two parallel straight lines;
- the space within a cylinder;
- the space between two coaxial cylinders;
- the space between two equidistant surfaces or two parallel planes;
- the space within a sphere.

Unless a more restrictive indication is required, for example by an explanatory note (see Figure 65), the toleranced feature may be of any form or orientation within this tolerance zone.

The tolerance applies to the whole extent of the considered feature unless otherwise specified.

Geometrical tolerances which are assigned to features related to a datum do not limit the form deviations of the datum feature itself. It may be necessary to specify tolerances of form for the datum feature(s).

2.4 Symbols

Symbols to indicate characteristics to be toleranced are shown in Table 1. The symbols in Table 2 identify and qualify toleranced features, datums, zones and dimensions. The uses of these symbols are shown in the remainder of this chapter.

Table 1 – Symbols for geometrical characteristics

Tolerances	Characteristics	Symbol	Datum needed	Subclause
Form	Straightness	—	no	18.1
Form	Flatness	▱	no	18.2
Form	Roundness	○	no	18.3
Form	Cylindricity	⌭	no	18.4
Form	Profile any line	⌒	no	18.5
Form	Profile any surface	⌓	no	18.7
Orientation	Parallelism	//	yes	18.9
Orientation	Perpendicularity	⊥	yes	18.10
Orientation	Angularity	∠	yes	18.11
Orientation	Profile any line	⌒	yes	18.6
Orientation	Profile any surface	⌓	yes	18.8
Location	Position	⌖	yes or no	18.12
Location	Concentricity (for centre points)	◎	yes	18.13
Location	Coaxiality (for axes)	◎	yes	18.13
Location	Symmetry	⌯	yes	18.14
Location	Profile any line	⌒	yes	18.6
Location	Profile any surface	⌓	yes	18.8
Run-out	Circular run-out	↗	yes	18.15
Run-out	Total run-out	⌰	yes	18.16

NOTE: The last column refers to clauses in BS EN ISO 1101.

Table 2 – Additional symbols

Description	Symbol	Reference
Toleranced feature indication		Clause 7
Datum feature indication	A , A	Clause 9 and ISO 5459
Datum target indication	⌀2 / A1	ISO 5459
Theoretically exact dimension	50	Clause 11
Median feature	Ⓐ	Clause 7
Projected tolerance zone	Ⓟ	Clause 13 and ISO 10578
Maximum material requirement	Ⓜ	Clause 14 and ISO 2692
Least material requirement	Ⓛ	Clause 15 and ISO 2692
Free state condition (non-rigid parts)	Ⓕ	Clause 16 and ISO 10579
All around (profile)		Subclause 10.1
Envelope requirement	Ⓔ	ISO 8015
Common zone	CZ	Subclause 8.5
Minor diameter	LD	Subclause 10.2
Major diameter	MD	Subclause 10.2
Pitch diameter	PD	Subclause 10.2
Line element	LE	Subclause 18.9.4
Not convex	NC	Subclause 6.3
Any cross-section	ACS	Subclause 18.13.1
Unequally disposed tolerance	UZ	Subclause 10.3
Intersection plane		Clause 19
Orientation plane		Clause 19
Collection plane		Clause 3
Direction feature		Clause 3
Between (two points)	↔	Subclause 10.1.4
From (two points)	→	Clause 8

NOTE: The last column refers to clauses in BS EN ISO 1101.

Geometric tolerancing datums and datum systems

2.5 Tolerance frame

The requirements are shown in a rectangular frame which is divided into two or more compartments. These compartments contain, from left to right, in the following order (as shown in Figures 44, 45, 46, 47 and 48):

- first compartment, the symbol for the geometrical characteristic;
- second compartment: information on the tolerance zone defined in the unit used for linear dimensions and complementary requirements. If the tolerance zone is circular or cylindrical, the value is preceded with the symbol 'Ø', if the tolerance zone is spherical, the value is preceded with 'SØ';
- third and subsequent compartment, if applicable: the letter or letters identifying the datum or common datum or datum system, as shown in Figures 45, 46, 47 and 48.

Figure 44 *Figure 45* *Figure 46* *Figure 47* *Figure 48*

When a tolerance applies to more than one feature, this shall be indicated above the tolerance frame by the number of features followed by the symbol 'x', as shown in Figures 49 and 50.

If required, indications qualifying the form of the feature within the tolerance zone shall be written near the tolerance frame, see Figure 51 and Table 2 for other indications.

If it is necessary to specify more than one geometrical characteristic for a feature, the requirements may be given in tolerance frames one under the other for convenience, as shown in Figure 52.

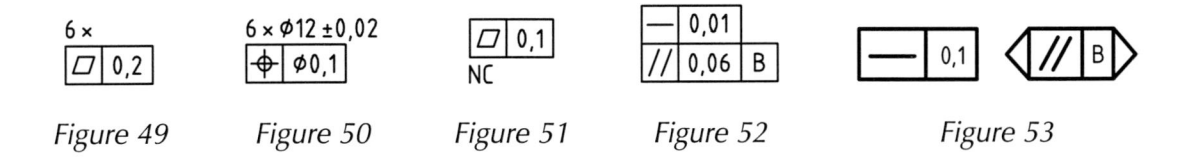

Figure 49 *Figure 50* *Figure 51* *Figure 52* *Figure 53*

If required, indications qualifying the direction of the tolerance zone and/or the extracted (actual) line shall be written after the tolerance frame. See Figure 53 for an example of indication of orientation of the tolerance zone.

2.6 Toleranced features

A geometrical specification tolerance applies to a single complete feature, unless an appropriate modifier is indicated. When the toleranced feature is not a single complete feature, see 2.9 Supplementary indications.

When the tolerance refers to the feature itself, the tolerance frame shall be connected to the toleranced feature by a leader line starting from either side of the frame and terminating with an arrowhead in one of the following ways:

- in 2D annotation, on the outline of the feature or an extension of the outline (but clearly separated from the dimension line), as shown in Figures 54 and 56; the arrowhead may also be placed on a reference line using a leader line to point to the surface, as shown in Figure 58.

- in 3D annotation, on the feature itself or on an extension line in continuation of the feature (but clearly separated from the dimension line), as shown in Figures 55 and 57; the arrowhead may also be placed on a reference line using a leader line to point to the surface, as shown in Figure 59.

NOTE: Leader lines are terminated with an arrow when they terminate on an outline of a feature, as shown in Figures 58 and 59, or a dot when they terminate on a surface, as shown in Figure 59. When the surface is visible, the dot is filled in, when the surface is hidden; the dot is not filled in.

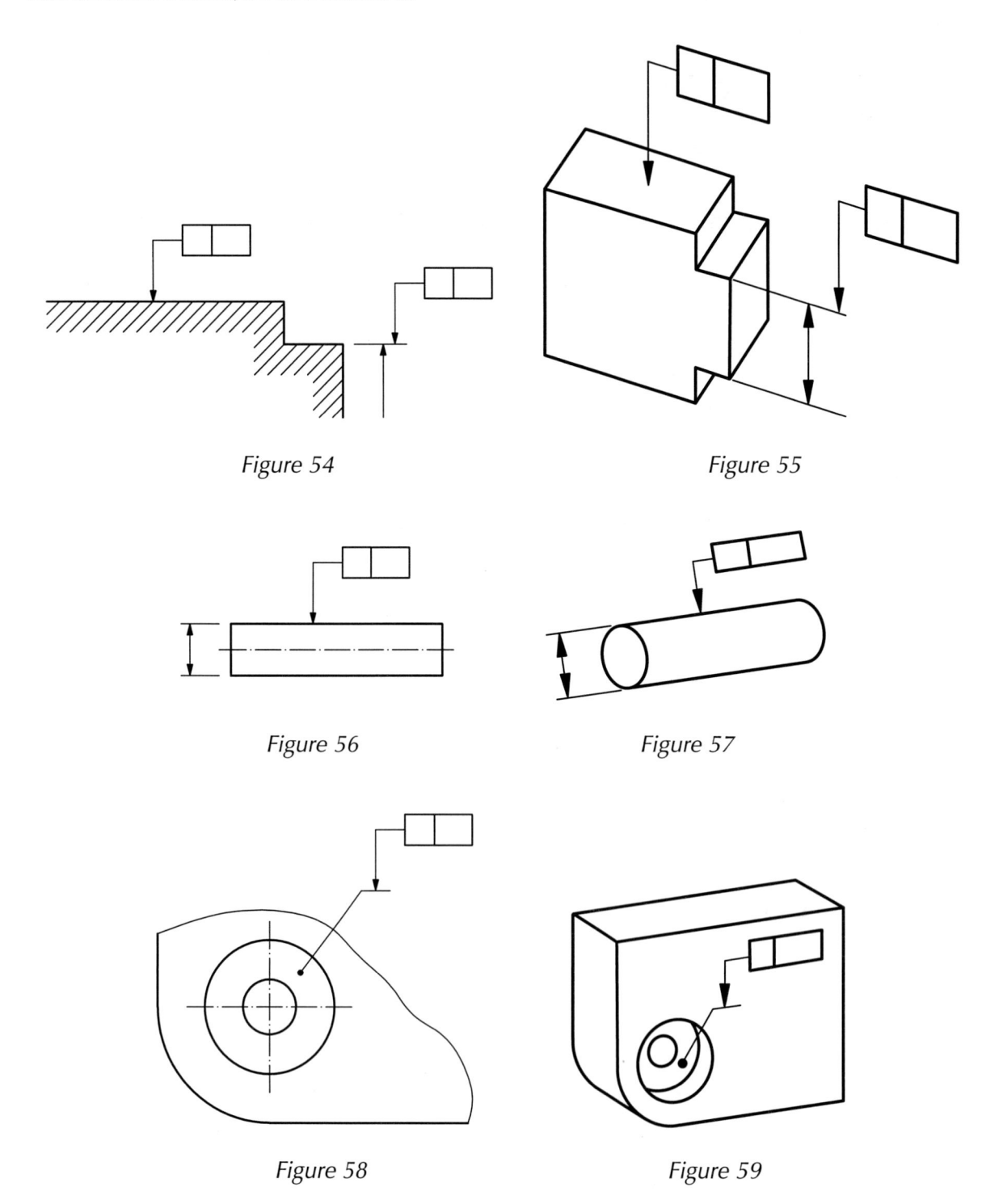

Figure 54

Figure 55

Figure 56

Figure 57

Figure 58

Figure 59

Geometric tolerancing datums and datum systems

When the tolerance refers to a median line, a median surface, or a median point, then it is indicated (in both 2D and 3D applications) either:

- by the leader line starting from the tolerance frame terminating on the extension of the dimension line of a feature of size, as shown in Figures 60, 61, 62, 63 and 64; or
- by a modifier Ⓐ (median feature) placed at the rightmost end of the second compartment of the tolerance frame (from the left). In this case, the leader line starting from the tolerance frame does not have to terminate on the dimension line, but can terminate on the outline of the feature or an extension of the outline, as shown in Figure 65.

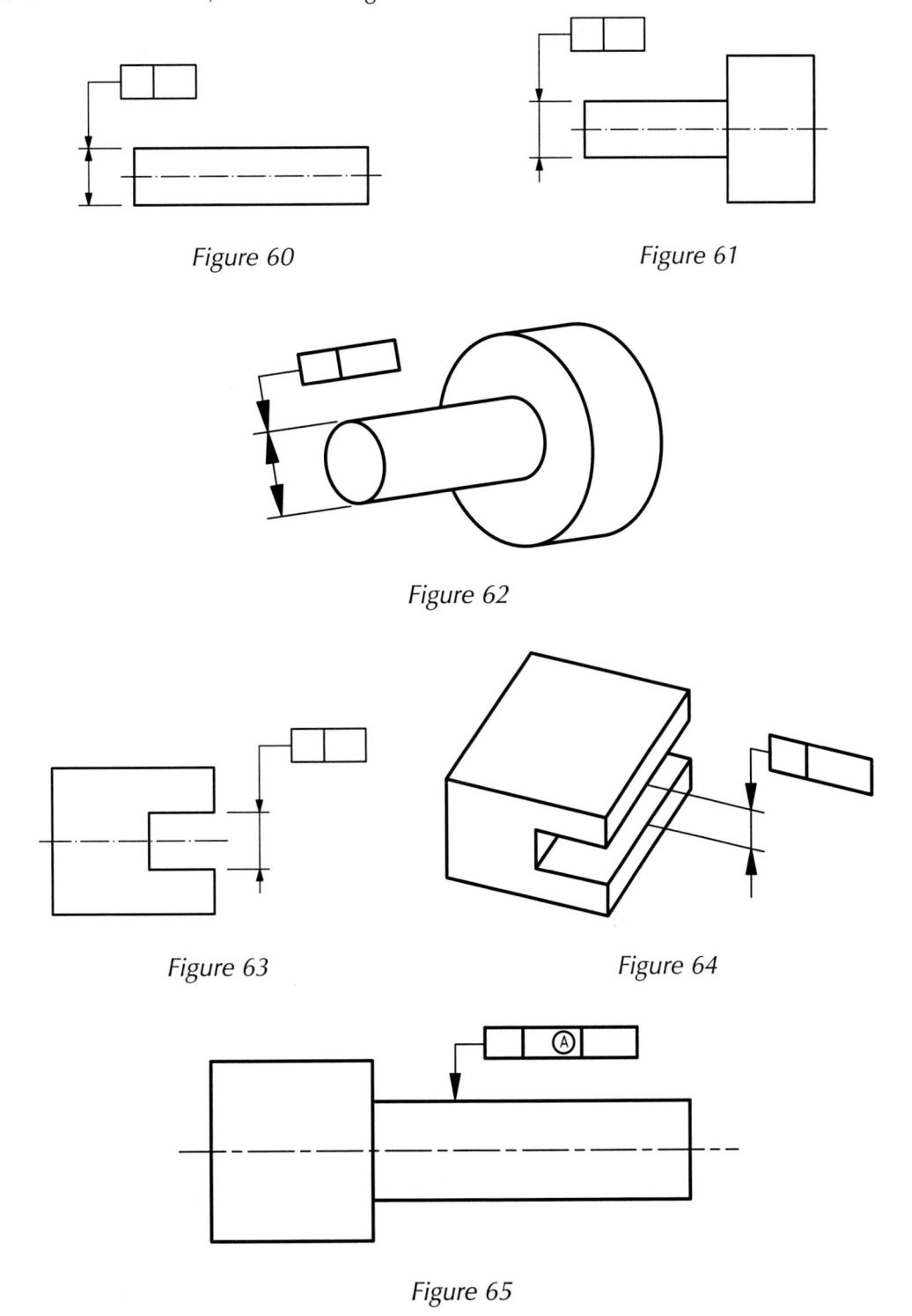

Figure 60

Figure 61

Figure 62

Figure 63

Figure 64

Figure 65

When the toleranced feature is a line, a further indication may be needed to control the orientation of the toleranced feature, see Figure 66a for 2D annotation and Figure 66b for 3D annotation.

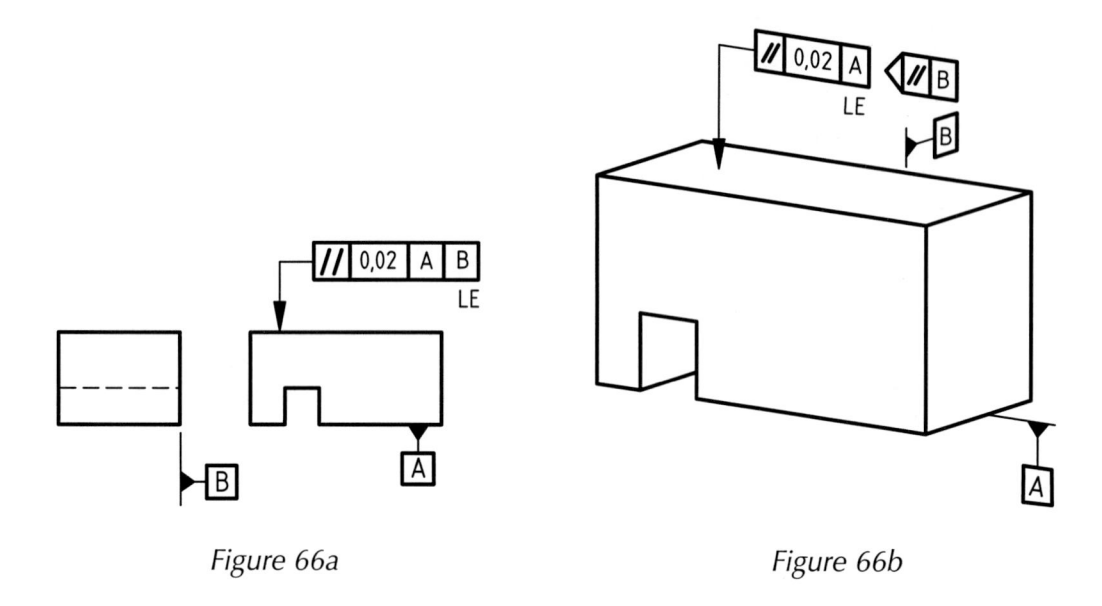

Figure 66a   *Figure 66b*

2.7 Tolerance zones

The tolerance zone is positioned symmetrically from the exact geometrical form, orientation, or location, unless otherwise indicated.

The tolerance value defines the width of the tolerance zone. This width applies normal to the specified geometry (see Figures 67 and 68) unless otherwise indicated (see figures 69 and 70).

NOTE: The orientation alone of the leader line does not influence the definition of the tolerance.

[a] Datum A.

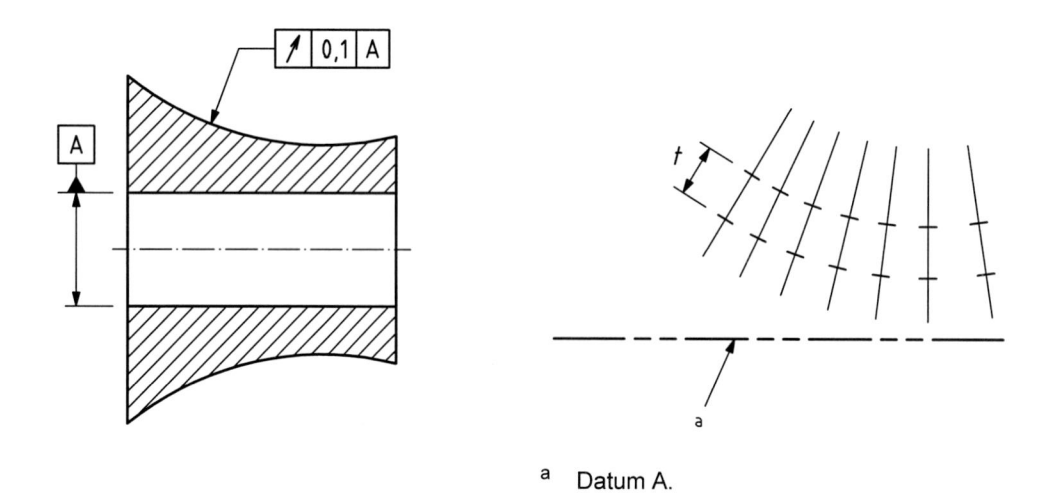

Figure 67 – Drawing indication   *Figure 68 – Interpretation*

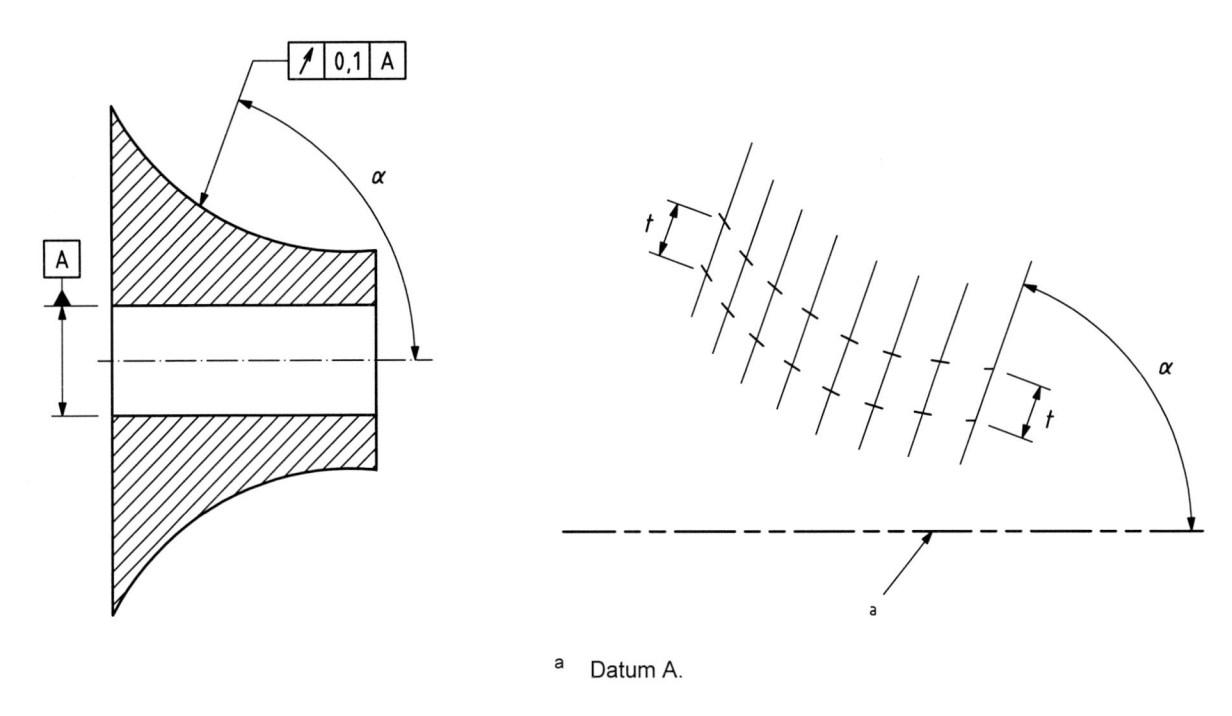

Geometric tolerancing datums and datum systems

Figure 69 – Drawing indication

a Datum A.

Figure 70 – Interpretation

The angle 'α' shown in Figure 69 shall be indicated, even if it is equal to 90°.

In the case of roundness, the width of the tolerance zone always applies in a plane perpendicular to the nominal axis.

The tolerance value is constant along the length of the considered feature, unless otherwise indicated either by:

- a graphical indication, defining a proportional variation from one value to another, between two specified locations on the considered feature, identified as given in clause 10.1.4. The letters identifying the locations are separated by an arrow; see Figure 71 and clause 12.2 for restricted parts of a feature. The values are related to the specified locations on the considered feature by the letters indicated over the tolerance frame (e.g. in Figure 71, the value of the tolerance is 0,1 for the location J and 0,2 for the location K). By default, the proportional variation follows the curvilinear coordinates, i.e. the distance along the curve connecting the two specified locations;
- a defined company specific indication, when the variation is not proportional.

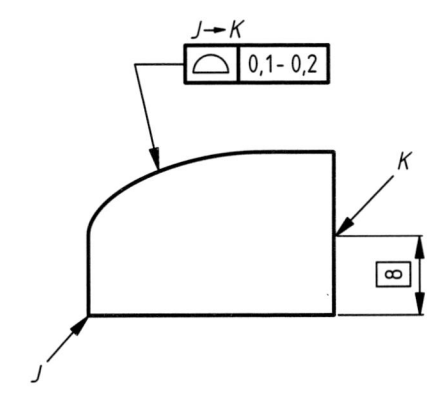

Figure 71

In the case of a centre point or median line or median surface toleranced in one direction:

- the orientation of the width of a positional tolerance zone is based on the pattern of the theoretically exact dimensions (TED) and is at 0° or 90° as indicated by the direction of the arrowhead of the leader line unless otherwise indicated, see Figure 72;
- the orientation of the width of an orientation tolerance zone is at 0° or 90° relative to the datum as indicated by the direction of the arrowhead of the leader line unless otherwise indicated, see Figures 73 and 74;
- when two tolerances are stated, they shall be perpendicular to each other unless otherwise specified, see Figures 73 and 74.

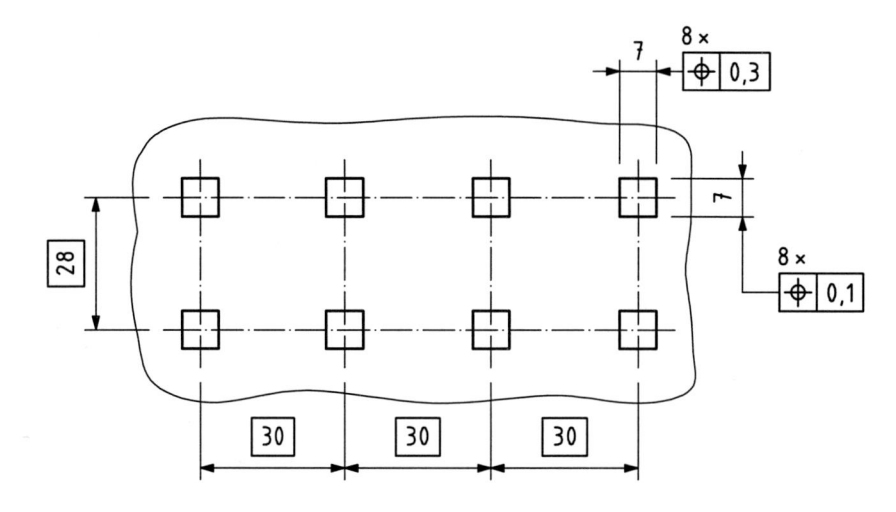

Figure 72

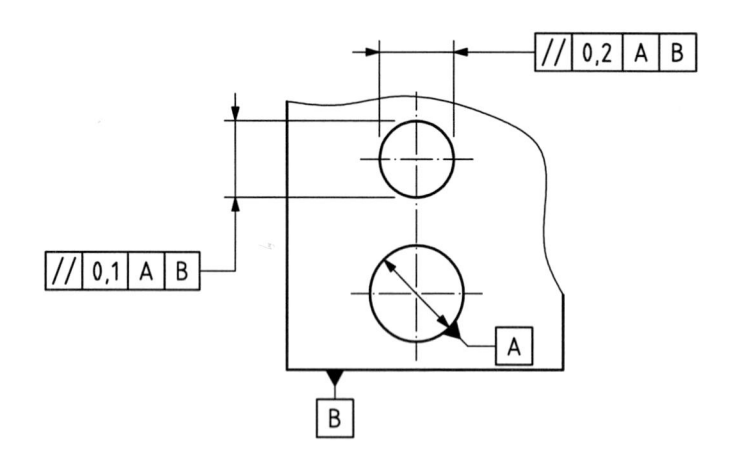

Figure 73 – Drawing indication

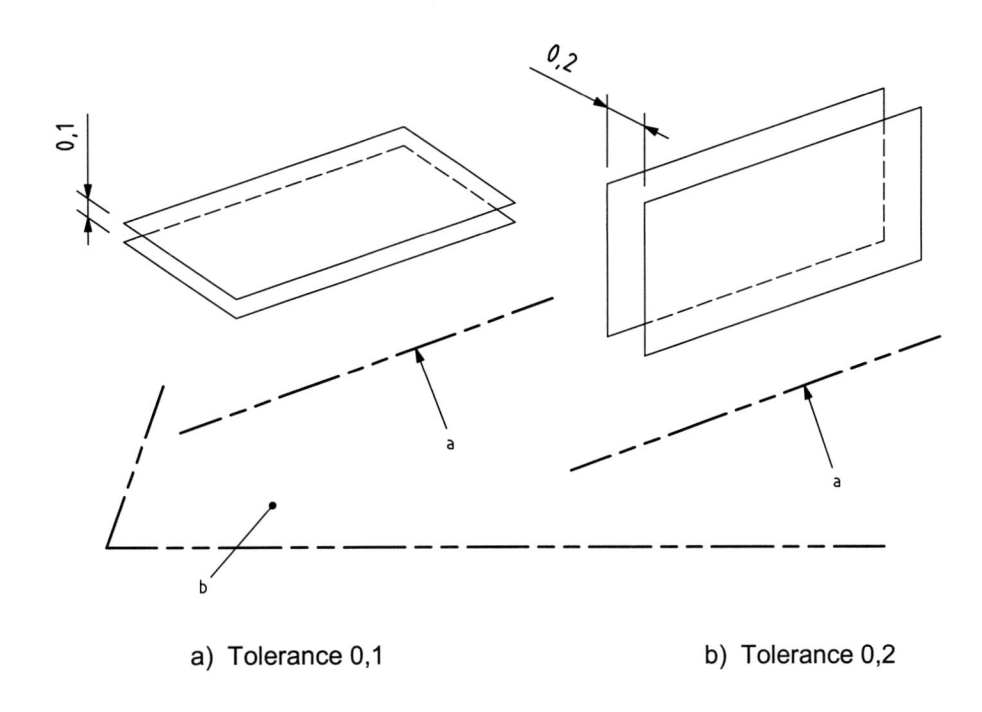

Geometric tolerancing datums and datum systems

a) Tolerance 0,1 b) Tolerance 0,2

a Datum A
b Datum B

Figure 74 – Interpretation

The tolerance zone is cylindrical or circular if the tolerance value is preceded by the symbol 'Ø' or spherical if it is preceded by the symbol 'SØ', as shown in Figures 75 and 76.

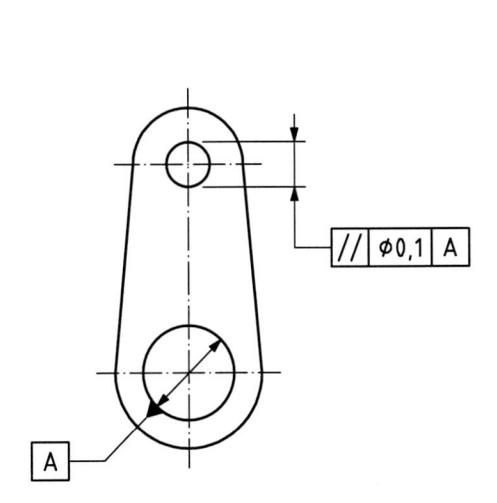

Figure 75 – Drawing indication 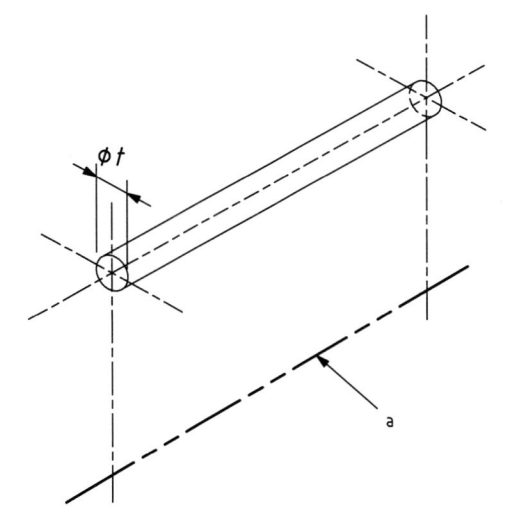 *Figure 76 – Interpretation*

Individual tolerance zones of the same value applied to several separate features may be specified as shown in Figure 77.

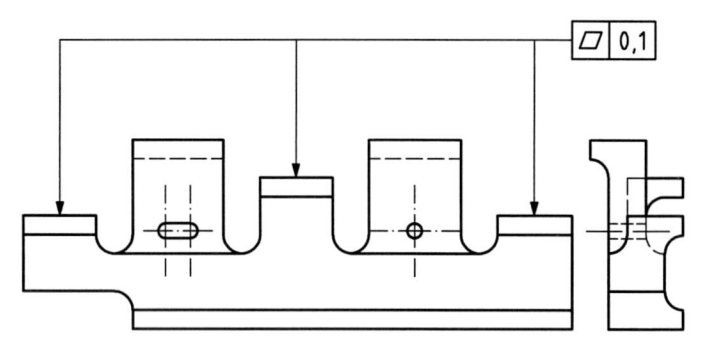

Figure 77

Where a common tolerance zone is applied to several separate features, this common requirement shall be indicated by the symbol 'CZ' for common zone following the tolerance in the tolerance frame, as shown in Figure 78.

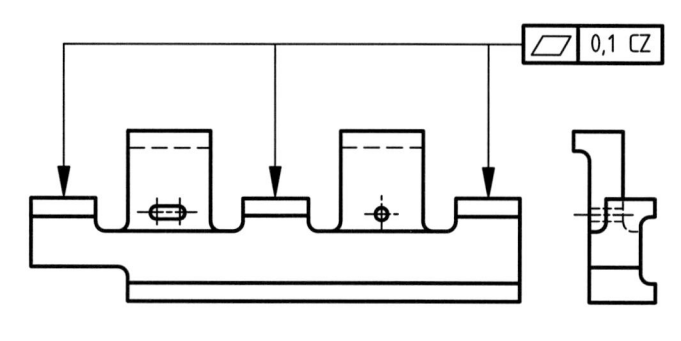

Figure 78

Where several tolerance zones (controlled by the same tolerance frame) are applied simultaneously (not independently) to several separate features, to create a combined zone, the requirement shall be indicated by the symbol 'CZ' for common zone following the tolerance in the tolerance frame, as shown in Figure 79. In addition, there shall be an indication that the specification applies to several features (e.g. using '3 x' over the tolerance frame as shown in Figures 49 and 50, or using leader lines attached to the tolerance frame as shown in Figure 77.

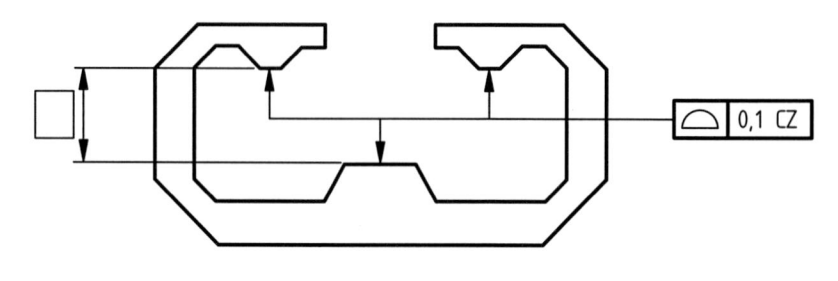

Figure 79

Where CZ is indicated in the tolerance frame, all the related individual tolerance zones shall be located and orientated amongst themselves using either implicit (0 mm, 0°, 90°, etc) or explicit theoretically exact dimensions (TED).

2.8 Datums and datum systems

Geometrical specifications define tolerancing of geometrical features. Tolerancing limits their geometrical deviations in relation to their theoretically exact form, orientation and/or location. This limitation is achieved by defining tolerance zones which confine the toleranced features of workpieces which conform to the specification. Depending on the application of these tolerance zones, three cases can exist:

- a tolerance zone is free to orient and locate itself to best accommodate the feature it confines, i.e. a form tolerance;
- a set of tolerance zones is oriented and located collectively, i.e. common zone;
- a tolerance zone is oriented and/or located in relation to other features, i.e. a datum or a datum system (see Examples 5 and 6).

Datums are established from identified real surfaces of the workpiece.

Datums can lock some degrees of freedom of a tolerance zone. The number of degrees of freedom locked (up to six) depends on the nominal shape of the features utilized to establish the datum or datum system and the toleranced characteristic indicated in the considered geometrical tolerance frame.

EXAMPLE 5: The tolerance zone, which is the space between two parallel planes 0,2 mm apart, is constrained in orientation by a 75° angle from the datum. Here, the datum is the cylinder axis. See Figure 80.

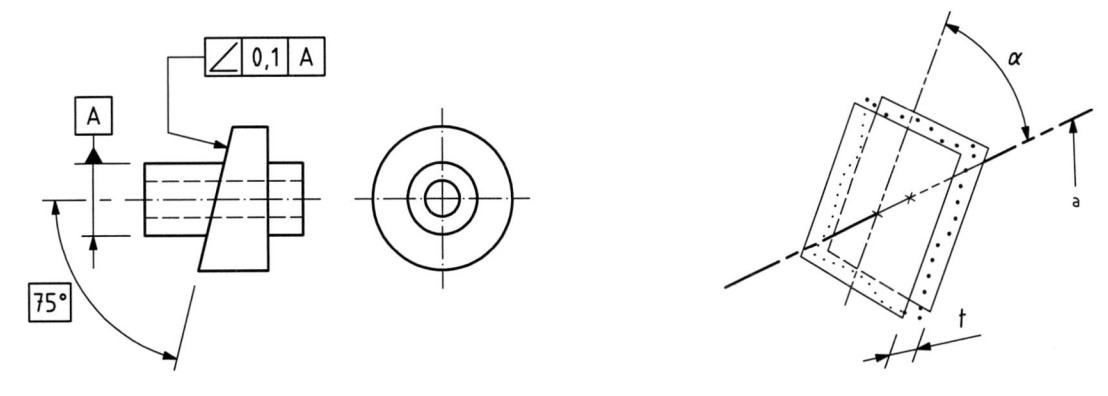

ᵃ   Datum A.

Figure 80 – Tolerance zone constrained in orientation by a 75° angle from the datum

EXAMPLE 6: The tolerance zone, which is the space between two parallel planes 0,2 mm apart, is constrained in orientation by a 110° angle from a datum, and in location by the distance 20 mm from the gauge plane (the plane where the local diameter of the cone with a fixed angle of 40°, is 30 mm). Here, the datum is the set of situation features of the cone with a fixed angle of 40°, i.e. the cone axis and the point of intersection between the gauge plane and that axis. See Figure 81.

ᵃ   Datum A.

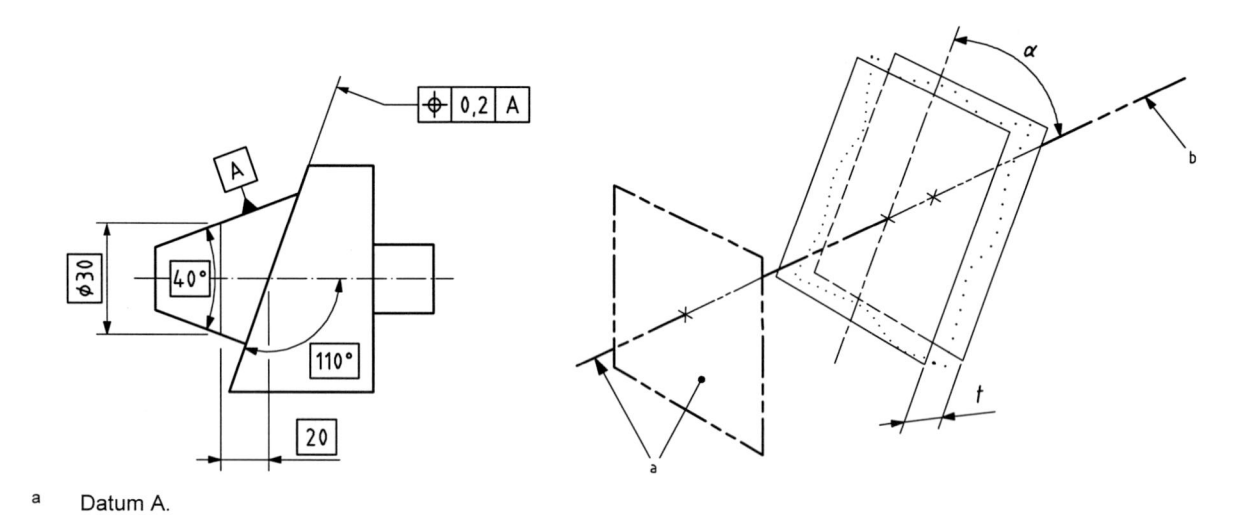

Figure 81 – Tolerance zone constrained in orientation by a 110° angle from the datum

A datum related to a toleranced feature shall be designated by a datum letter. A capital letter shall be enclosed in a datum frame and connected to a filled or open datum triangle to identify the datum, see Figures 82 and 83 for 2D application and Figure 84 for 3D application; the same letter which defines the datum shall also be indicated in the tolerance frame.

It is recommended not to use the letters I, O, Q and X which can be misinterpreted.

There is no difference in the meaning between a filled and an open datum triangle.

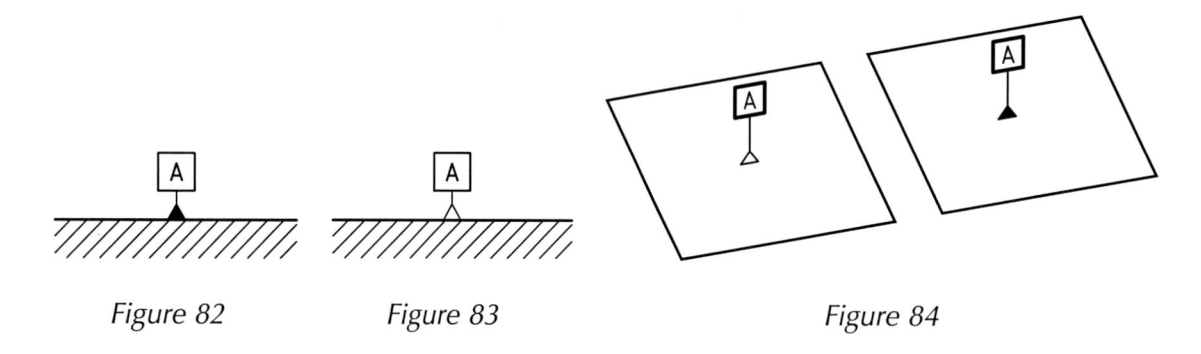

Figure 82   *Figure 83*   *Figure 84*

When a datum is a feature (such as a surface), the datum triangle with the datum letter shall be placed:

- in 2D annotation, on the outline of the feature or an extension of the outline (but clearly separated from the dimension line), when the datum is the line or surface shown, see Figure 85; the datum triangle may also be placed on a reference line using a leader line to point to the surface, as shown in Figure 86;
- In 3D annotation, on the feature itself or on an extension line in continuation of the feature (but clearly separated from the dimension line) when the datum is the line or surface shown, see Figure 87; the datum triangle may be placed on a reference line using a leader line to point to the surface, as shown in Figure 88.

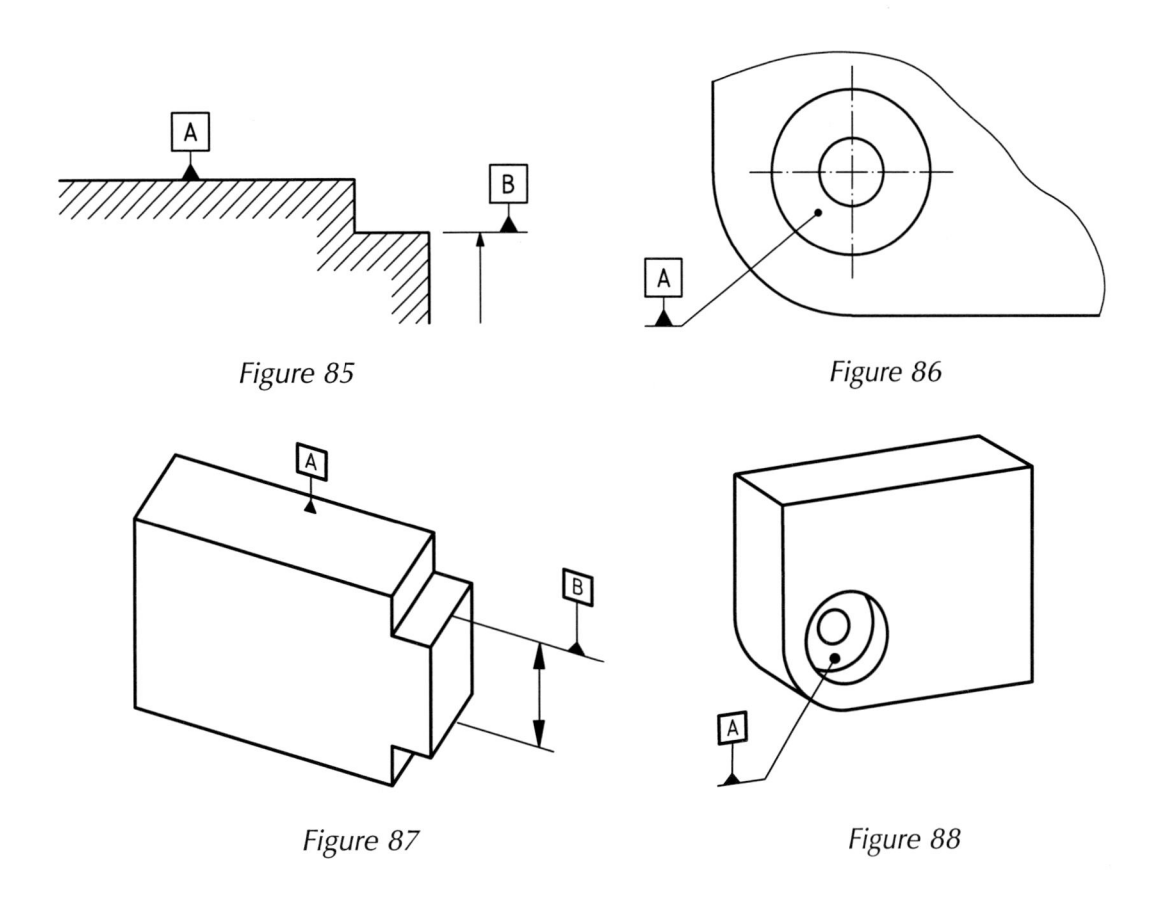

Geometric tolerancing datums and datum systems

Figure 85

Figure 86

Figure 87

Figure 88

When the datum is the axis or median plane or a point defined by the feature so dimensioned, the datum triangle with the datum letter shall be placed as an extension of the dimension line in both 2D and 3D applications, as shown in Figures 89 to 94. If there is insufficient space for two arrowheads, one of them may be replaced by the datum triangle, as shown in Figures 91 to 94.

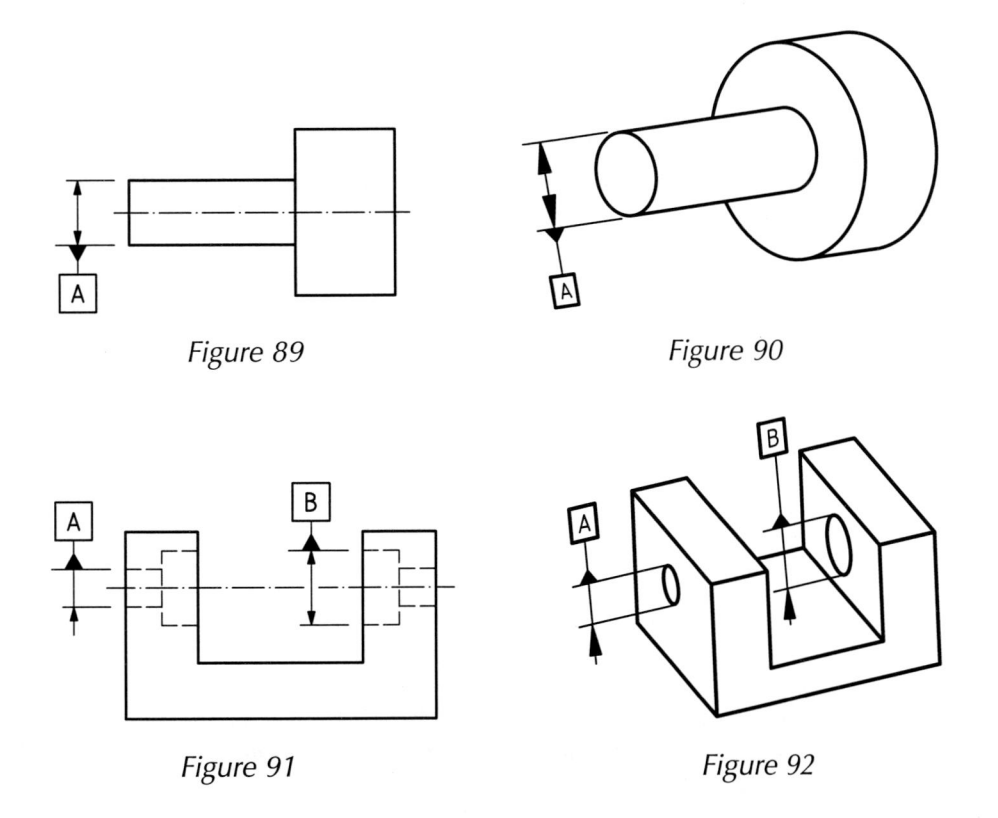

Figure 89

Figure 90

Figure 91

Figure 92

Figure 93   Figure 94

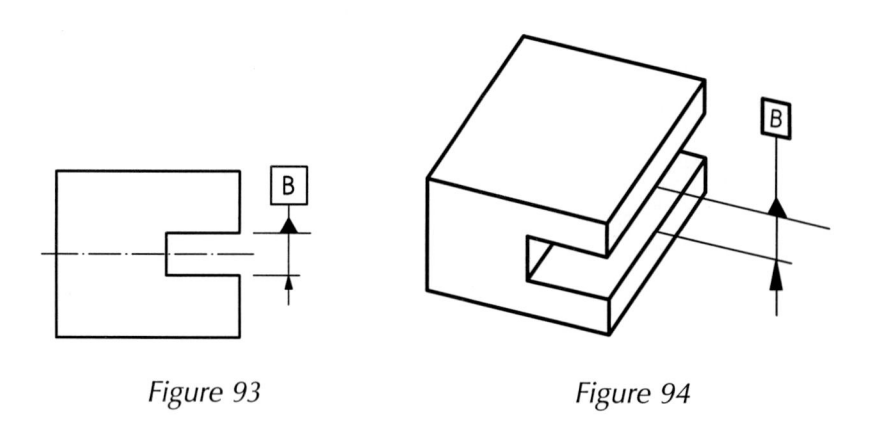

If a datum is applied to a restricted part of a feature only, this restriction shall be shown as a wide, long dashed-dotted line and dimensioned using TEDs, as shown in Figures 95 and 96. (See BS ISO 128-24, Table 2, 04.2 for line type.)

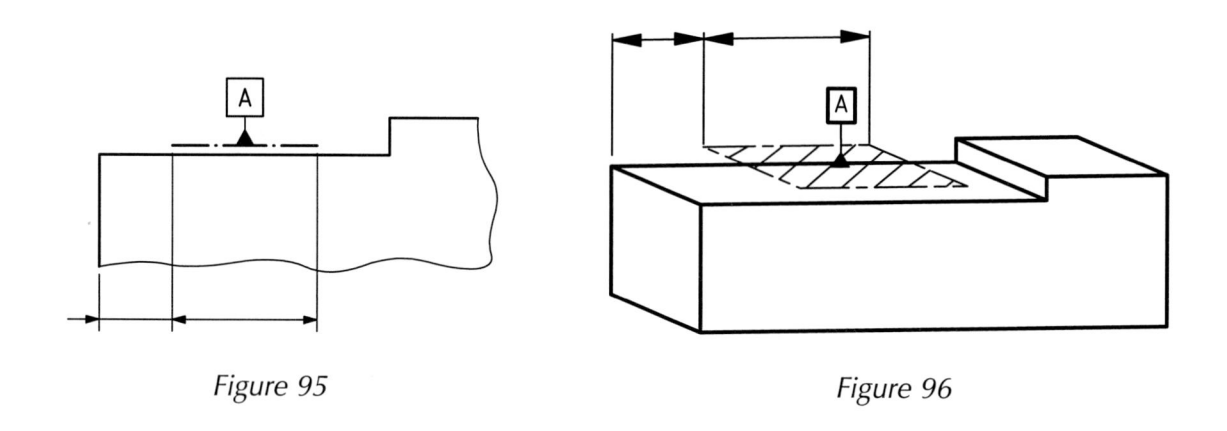

Figure 95   Figure 96

Tolerance frames with only three compartments signify a single datum or a common datum used alone; see Figures 97a and 97b.

When a datum system is specified, the tolerance frame shall have more than three compartments, as shown in Figure 97c to 97e.

Each compartment of the tolerance frame (after the second) shall contain either a single datum or a common datum.

In a datum system, the primary datum is identified in the third compartment of the tolerance frame; the secondary datum is identified in the fourth compartment of the tolerance frame; the tertiary datum is identified in the fifth compartment of the tolerance frame, as shown in Figures 97c, d and e.

When a datum system is used, the orientation constraints between each datum (single or common) are specified by TEDs. TED values of 0°, 90°, 180° and 270° are implicit and not indicated.

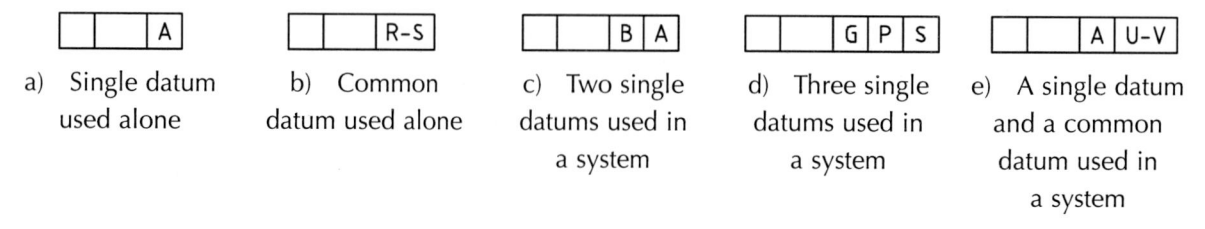

a) Single datum used alone
b) Common datum used alone
c) Two single datums used in a system
d) Three single datums used in a system
e) A single datum and a common datum used in a system

Figure 97 – Examples of indication of datums in the tolerance frame

Datums can also be identified by either:

- adding the complementary indication 'n' giving the number (n) of surfaces in the collection on the right side of a datum indicator attached to one of the surfaces, as shown in Figure 98a. When the datum indicator points to the tolerance frame, the indication 'n' is not written on the right side of a datum indicator but above the tolerance frame, as shown in Figure 98b, or,
- when the datum indicator points to the tolerance frame, by using leader lines indicating each surface included in the common datum, as shown in Figure 98c.

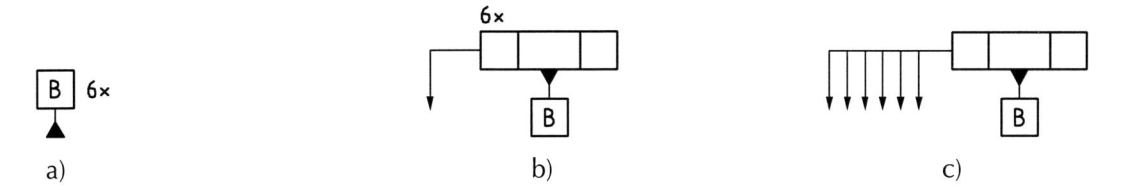

Figure 98 – Examples complementary indication of datums

2.8.1 Datum targets

When it is not desirable to use a complete integral surface to establish a datum feature, it is possible to indicate portions of the surface (areas, lines or points) and their dimensions and locations. These portions are called datum targets. They usually simulate the interface between the portion of the considered surface of the workpiece and one or more contacting ideal features (assembly interface features or fixture features).

A datum target is indicated by a datum target indicator. This indicator is constructed from a datum target frame, a datum target symbol and a leader line linking the two symbols (directly, or through a reference line).

Where the datum target indicator uses a single datum target frame, the datum target is the portion of the integral surface (point, line or area) indicated.

Where the datum target indicator uses an equalizing datum target frame, the datum target is the median feature constructed from two portions (a pair) of the same type (point, line or area) of the integral surface indicated, and the distance between these two portions is variable.

The datum target frame is divided into two compartments by a horizontal line. The lower compartment is reserved for a letter and a digit (from 1 to n). The letter represents the datum feature and the digit the datum target number.

The upper compartment is reserved for additional information, such as dimensions of the target area.

The single datum target frame is a circle, as shown in Figure 99. The equalizing datum target frame is a hexagon, as shown in Figure 100.

Figure 99 – Single datum target frame

Figure 100 – Equalizing datum frame

The types of datum targets are a point, a line and an area. They are indicated using the following datum target symbols, respectively:

- a cross, as shown in Figure 101a;
- a long-dashed double-dotted narrow line (type 05.1 of BS ISO 128-24), which, when this line is not closed, is terminated by two crosses, as shown in Figures 101b and c. This line may be straight, circular or of any shape;
- a hatched area surrounded by a long-dashed double-dotted narrow line (type 05.1 of BS ISO 128-24), as shown in Figure 101d and e.

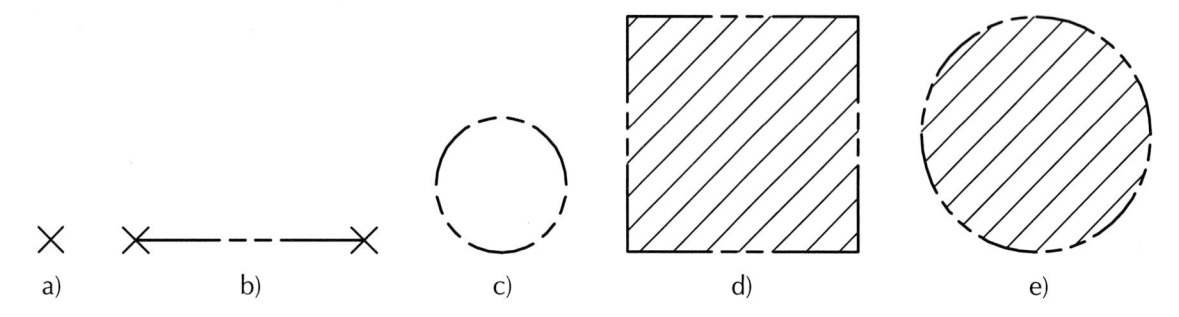

Figure 101 – Datum target indication

The datum target frame is connected directly, or through a reference line, to the datum target symbol by a leader line terminated with an arrow, as shown in Figures 102a and b. When the portion of the surface in question is hidden, the hidden part of the leader line shall be dashed and terminated by an open circle.

When the leader line is used with a single datum target frame, the orientation of the leader line connecting the frame with the datum target symbol is unimportant.

The leader line shall indicate the direction of movement of the relevant datum target when used with an equalizing datum target frame. This direction shall be given by the segment of the leader line terminated by the arrow (or the open circle), in an appropriate view. Where the required direction of movement is ambiguous, the angle shall be indicated as a TED between the leader line and the surface of the workpiece.

NOTE: The complete definition of this direction may require more than one 2D view.

A datum target indicator is constructed from a datum target frame, a datum target symbol and a leader line linking the two symbols, directly, or through a reference line, as shown in Figures 102a, b and c.

It may be necessary to indicate the same datum target on several appropriate views to have an unambiguous definition of the considered datum target, as shown in Figure 107.

For equalizing datum target frames, the leader line shall be in the extension of a dimension line (with no TED value) linking the two datum target symbols, as shown in Figures 103 and 104. The indication of the second datum target symbol of the pair and the dimension line may be omitted when no ambiguity exists.

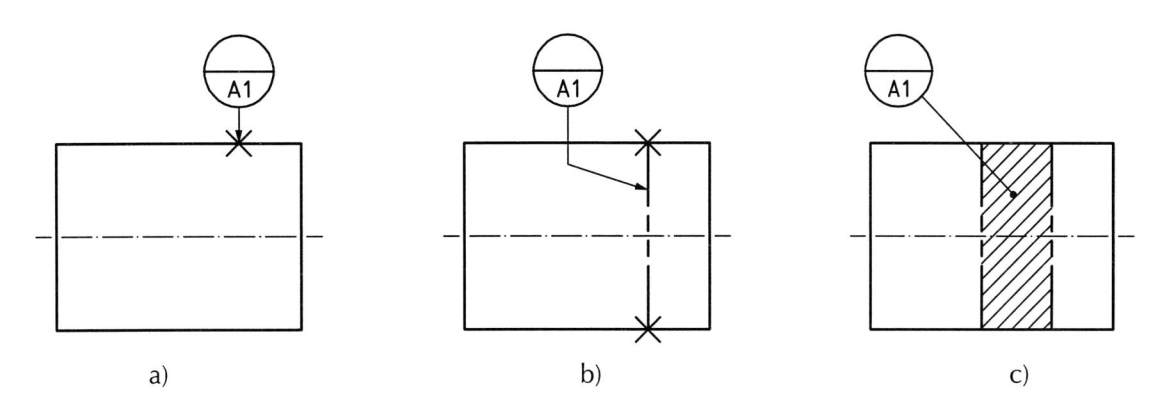

Geometric tolerancing datums and datum systems

a)   b)   c)

Figure 102 – Datum target indicators for a point, line and surface

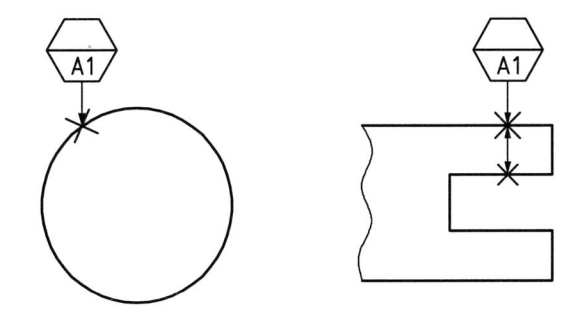

Figure 103 – Point

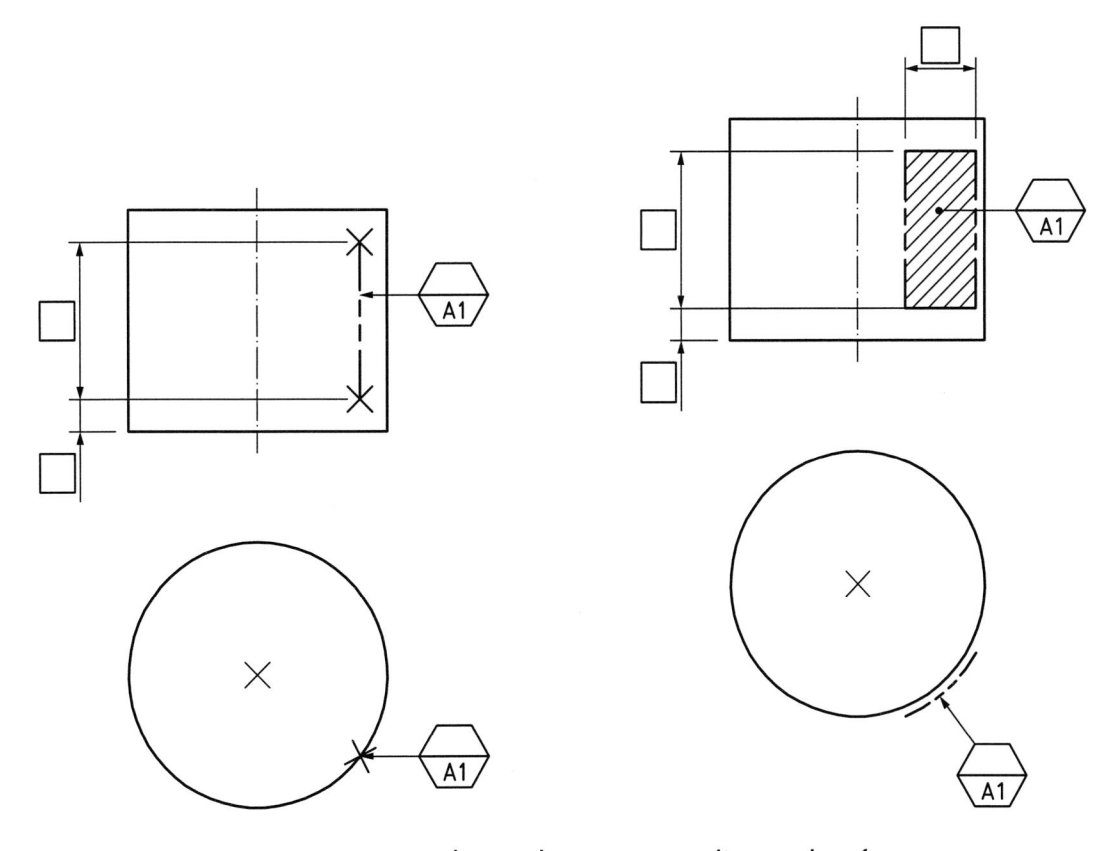

Figure 104 – Equalizing datum target – line and surface

If a datum is established from datum targets belonging to only one surface, then the letter identifying the surface shall be repeated on the right side of the datum indicator, followed by the list of numbers (separated by commas) identifying the targets, as shown in Figure 105. Each individual datum target shall be identified by a datum target indicator, indicating the datum letter, the number of the datum target and, if applicable, the dimensions of the datum target, as shown in Figure 107.

Figure 105 – Indication of datums established from datum targets

It is permitted to simplify the drawing indication in case of only one datum target by placing the datum indicator on a long-dashed dotted wide line (type 04.2 of BS ISO 128-24) defining the portion of the considered surface, as shown in Figure 106a or on the reference line of a leader line pointing to a hatched area surrounded by a long dashed double-dotted narrow line (type 05.1 of BS ISO 128-24), as shown in Figure 106b.

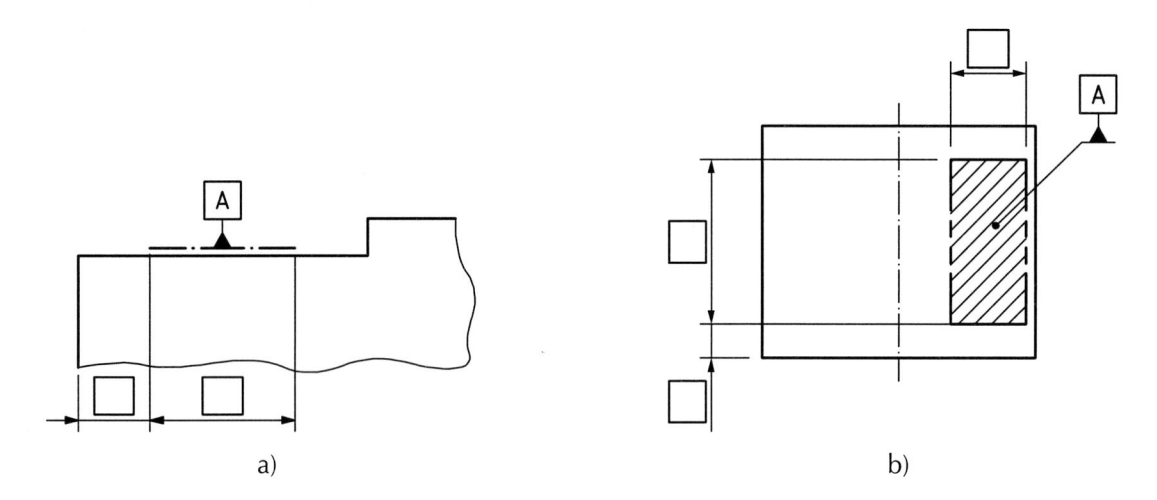

Figure 106 – Simplification of drawing indication in the case of only one datum target area

The location of datum targets on one surface shall be defined by TEDs, as shown in Figure 107. The location of a datum target relative to one or more other feature(s) shall also be defined by TEDs.
The length of a datum target line shall be defined by TEDs.
The extent of a datum target area shall be considered theoretically exact. The dimensions of the area shall be indicated:

- in the upper compartment of the datum target indicator when the area is circular or square, as shown in Figure 108, or, if the area is rectangular and the space within the compartment is limited, placed outside and connected to the appropriate compartment by either a leader line or by a leader line and a reference line, as shown in Figure 109;
- directly on the drawing by TEDs, when the area is neither square, circular nor rectangular.

NOTE: In case of a point or a line, it may be necessary to indicate the datum target on several views to have an unambiguous definition.

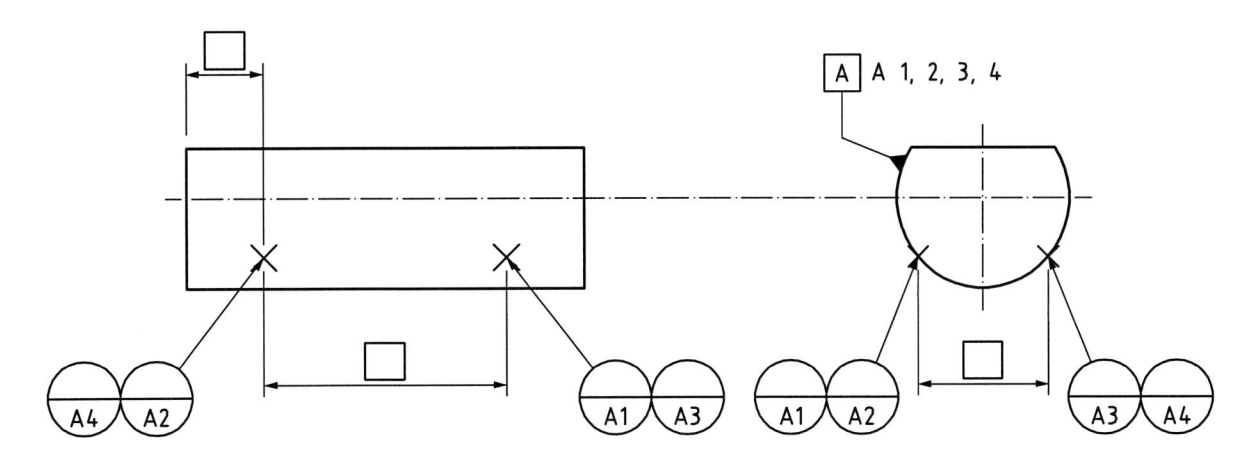

Geometric tolerancing datums and datum systems

Figure 107 – Datum targets on one surface

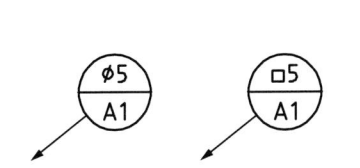

Figure 108 – Indication of circular or square area

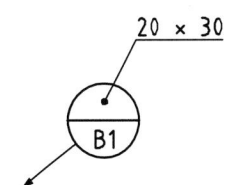

Figure 109 – Indication of rectangular area

2.9 Supplementary indications

When the toleranced feature is a portion of a single feature, or a compound continuous feature, then it shall be indicated either as a:

- continuous, closed feature (single or compound), or
- restricted area of a single surface, or
- continuous, non-closed feature (single or compound).

2.9.1 All around – Continuous, closed toleranced feature

If a requirement applies to a closed compound continuous surface defined by a collection plane, the 'all around' modifier ('O') shall be placed on the intersection of the leader line and the reference line of the tolerance frame. In addition, a collection plane indicator identifying the collection plane shall be placed after the tolerance frame, as shown in Figures 110 to 113. An all-around requirement applies only to the surfaces represented by the outline, not the entire workpiece, as shown in Figures 110 and 112.

If a requirement applies to the set of line elements on the closed compound continuous surface (defined by a collection plane), an intersection plane indicator identifying the intersection plane shall also be placed between the tolerance frame and the collection plane indicator, as shown in Figure 111.

NOTE: The long dashed short dashed line indicates the considered features. Surfaces a and b are not considered in the specification.

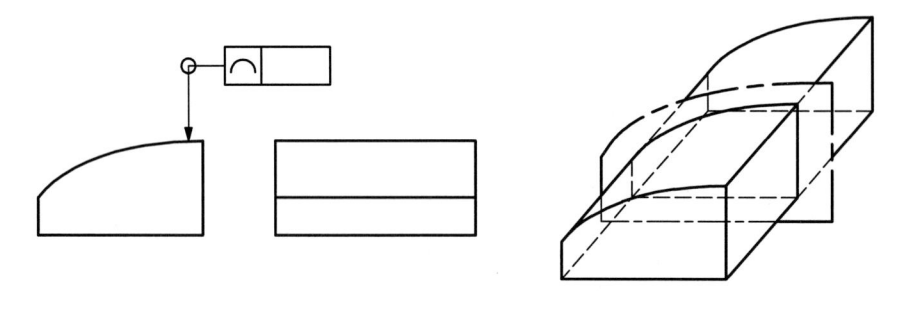

Figure 110

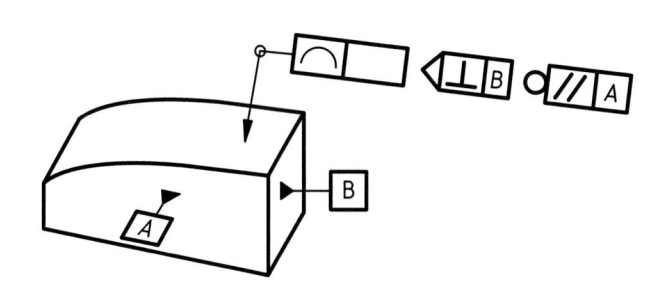

Figure 111

NOTE: When using the any line symbol, if the intersection plane and the collection plane are the same, the collection plane symbol can be omitted.

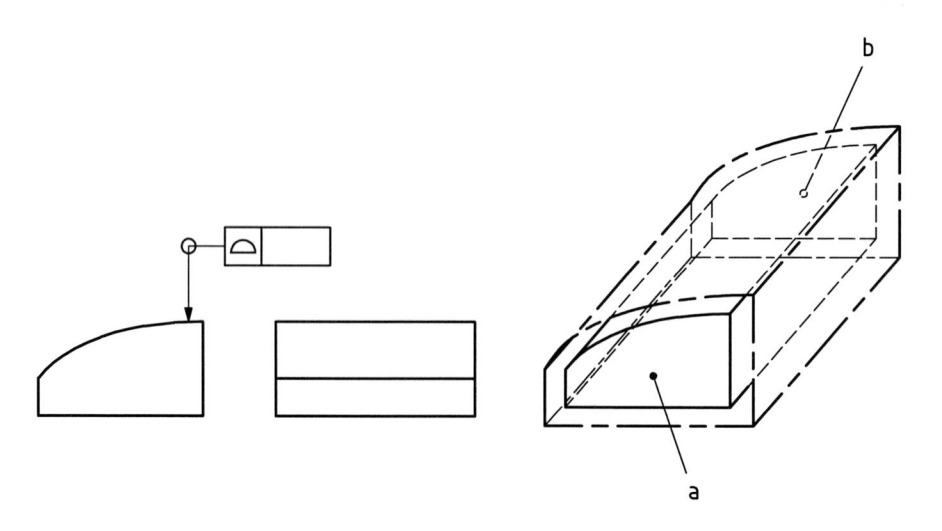

Figure 112

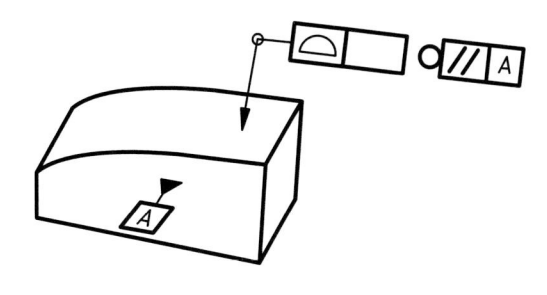

Geometric tolerancing datums and datum systems

Figure 113

2.9.2 Restricted area toleranced feature

In 2D annotation, the surface portions involved shall be outlined by a long-dashed dotted wide line (in accordance with BS ISO 128-24), as shown in Figures 114 and 115.

In 3D annotation, the leader line starting from the tolerance frame shall terminate on a hatched area, indicating the surface portions involved.

The location and dimensions of the surface portion shall be defined by TEDs, as shown in Figures 115 and 117.

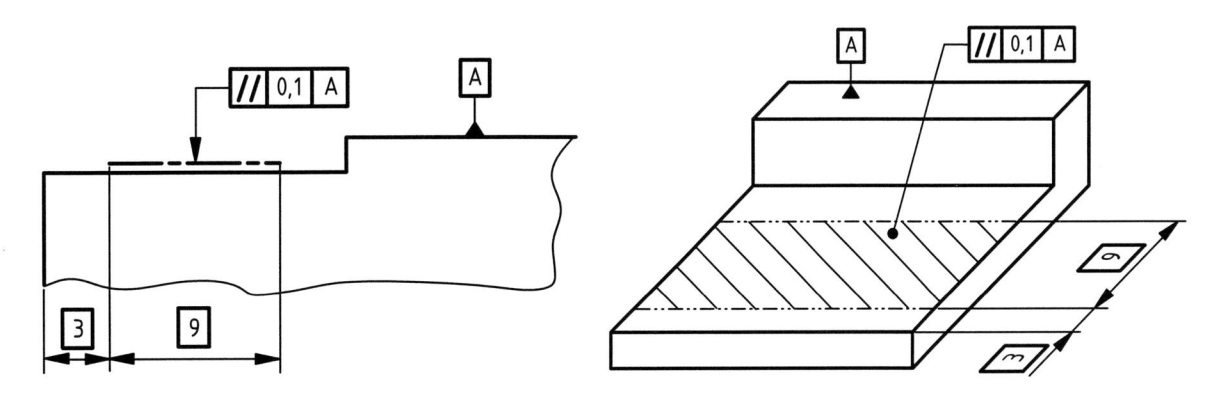

Figure 114

Figure 115

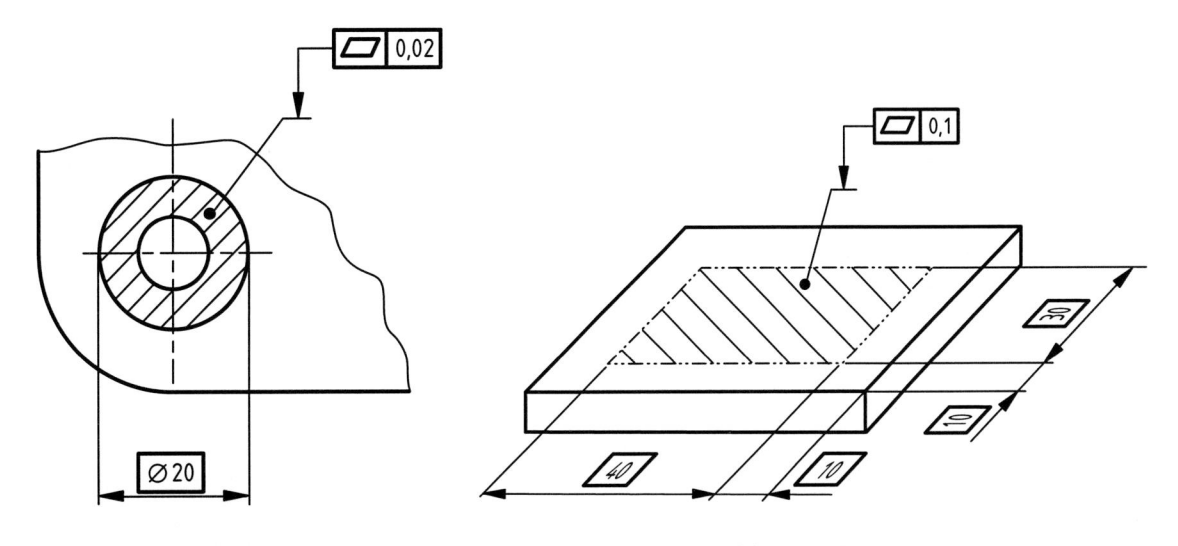

Figure 116

Figure 117

2.9.3 Continuous, non closed toleranced feature

If a tolerance applies to one identified restricted part of a feature or to contiguous restricted parts of contiguous features, but does not apply to the entire outline of the cross-sections (or entire surface represented by the outline), this restriction shall be indicated using the symbol '↔' (called 'between') and by identifying the start and the end of the toleranced feature.

The points or lines that identify the start and end of the toleranced feature are each identified by a capital letter connected to it by a leader line terminating with an arrowhead. If the point or line is not at the boundary of an integral feature, its location shall be indicated by TEDs.

The between symbol '↔' is used between two capital letters that identify the start and the end of the toleranced feature. This feature (compound toleranced feature) consists of all segments or areas between the start and the end of the identified features or parts of features.

In order to clearly identify the toleranced feature, the tolerance frame shall be connected to the compound toleranced feature by a leader line starting from either side of the frame and terminating with an arrowhead on the outline of the compound toleranced feature, as shown in Figure 117. The arrowhead may also be placed on a reference line using a leader line to point to the surface.

In Figure 118, the toleranced feature is the upper surface starting at line J and finishing at line K. The long dashed dotted line represents the toleranced feature, Surfaces a, b and c are not covered by the specification.

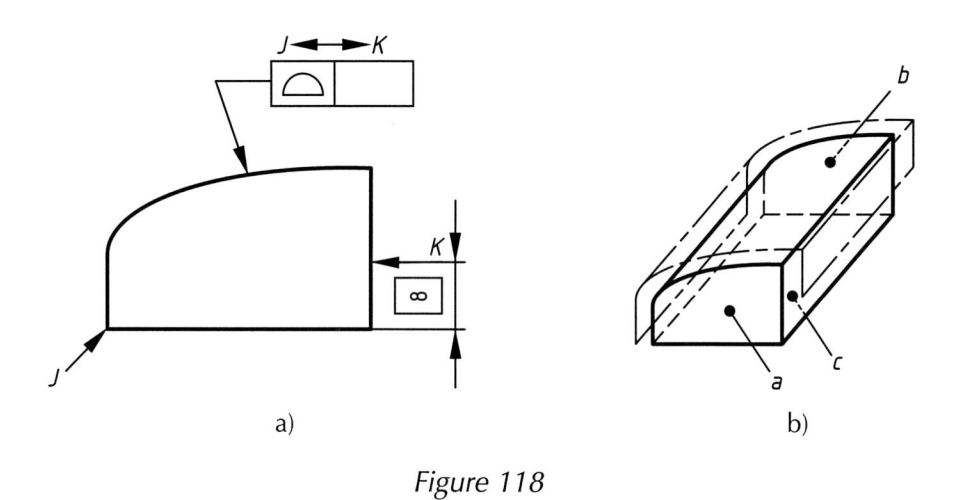

Figure 118

To avoid problems of interpretation regarding the considered feature, the start and end of the feature shall be indicated as shown in Figure 119.

If the same specification is applicable to a set of compound toleranced features, this set can be indicated above the tolerance frame, one above the other, as shown in Figure 120.

If all the compound toleranced features in the set are defined identically, it is possible to simplify the indication of this set, using the 'n x' indication, i.e. '3 x', as shown in Figure 121.

The rule defining the common zone indication also applies to defining a common compound tolerance zone, as shown in Figure 121.

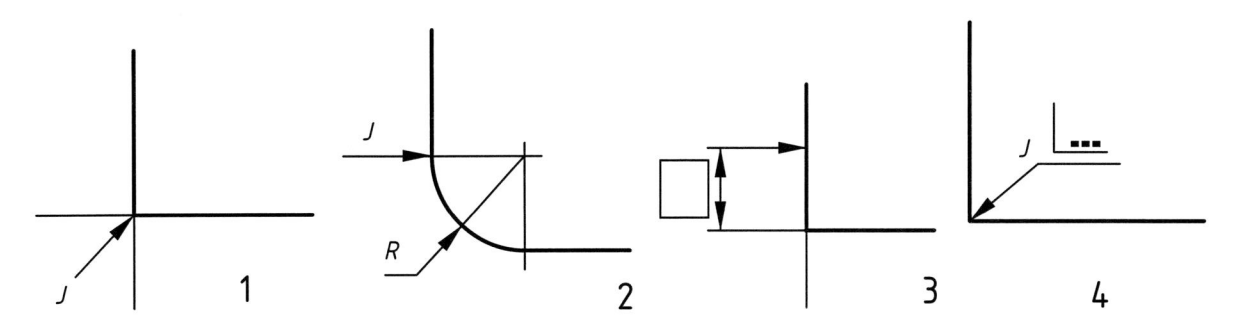

Geometric tolerancing datums and datum systems

Key
1. sharp edge or corner
2. rounded 'blind' edge (tangent continuity)
3. offset from corner or edge (with TED)
4. combination with an edge indication according to BS ISO 13715

Figure 119

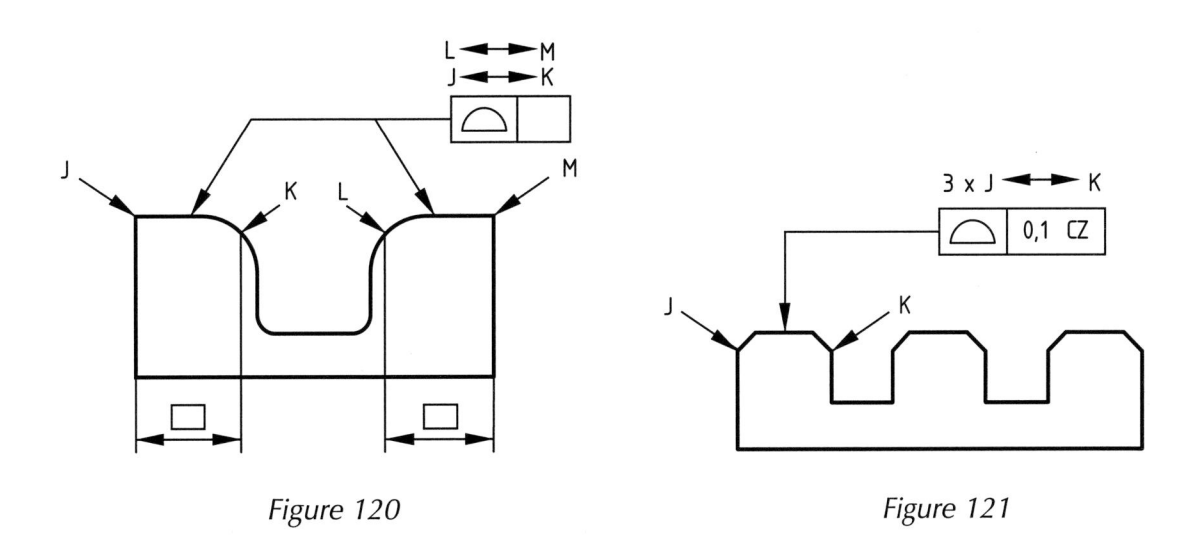

Figure 120

Figure 121

2.9.4 Unequally disposed tolerance zone

If the tolerance zone is not centred on the theoretically exact geometrical form, then this unequally disposed tolerance zone shall be indicated using the 'UZ' modifier as shown in Figure 122.

The extracted (actual) surface shall be contained between two equidistant surfaces enveloping spheres of a diameter equal to the tolerance value, the centres of which are situated on a surface corresponding to the envelope of a sphere in contact with the theoretically exact geometrical form and whose diameter is equal to the absolute value given after the 'UZ' modifier with the direction of the shift indicated by the sign, plus (+) indicating out of the material and minus (−) into the material.

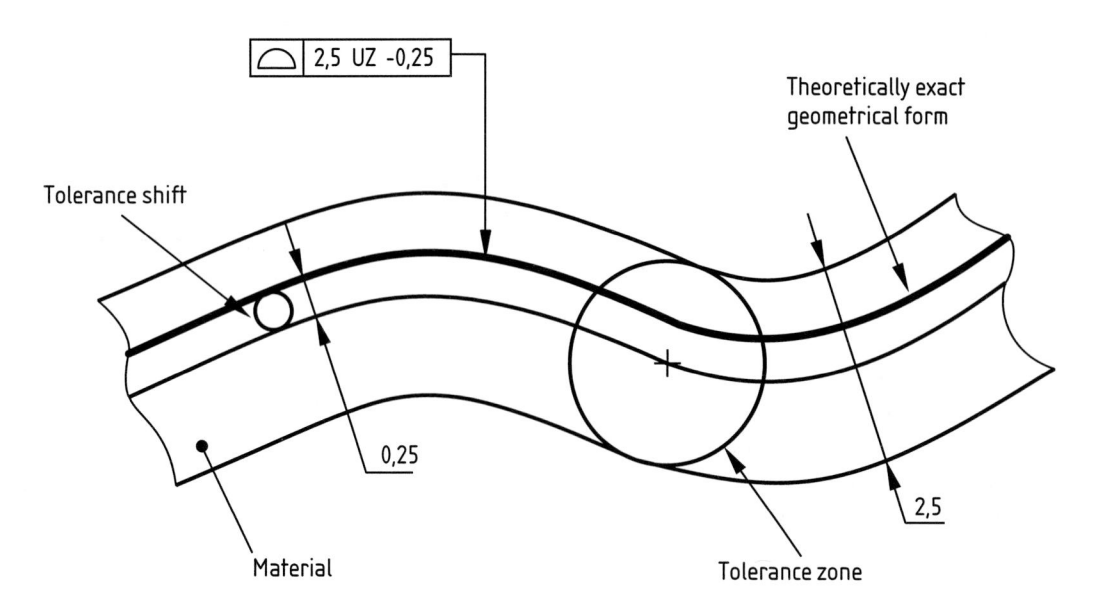

Figure 122 – Unequally disposed tolerance zone indication

When specifying a unilaterally disposed tolerance zone, the value after the 'UZ' modifier shall be equal to half the value of the tolerance zone.

2.9.5 Screw threads

Tolerances and datums specified for screw threads apply to the axis derived from the pitch cylinder, unless otherwise specified, e.g. 'MD' for major diameter and 'LD' for minor diameter, as shown in Figure 123. Tolerances and datums specified for gears and splines shall designate the specific feature to which they apply, i.e. 'PD' for pitch diameter, 'MD' for major diameter or 'LD' for minor diameter.

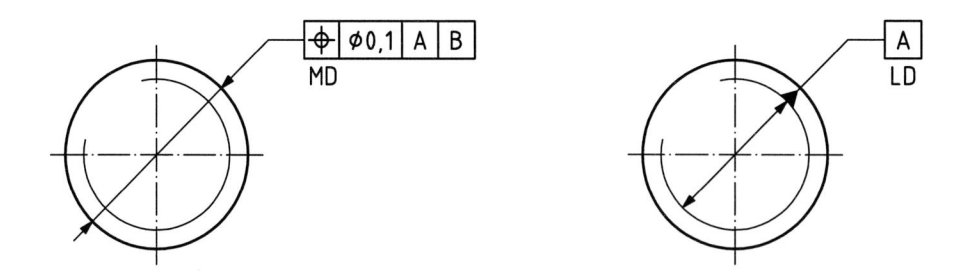

Figure 123

2.9.6 Theoretically exact dimensions (TED)

If tolerances of location, orientation or profile are prescribed for a feature or a group of features, the dimensions determining the theoretically exact location, orientation or profile respectively are called theoretically exact dimensions.

TED also applies to the dimensions determining the relative orientation of the datums of a system. TEDs shall not be toleranced. They are to be enclosed in a frame, as shown in Figures 124 and 125.

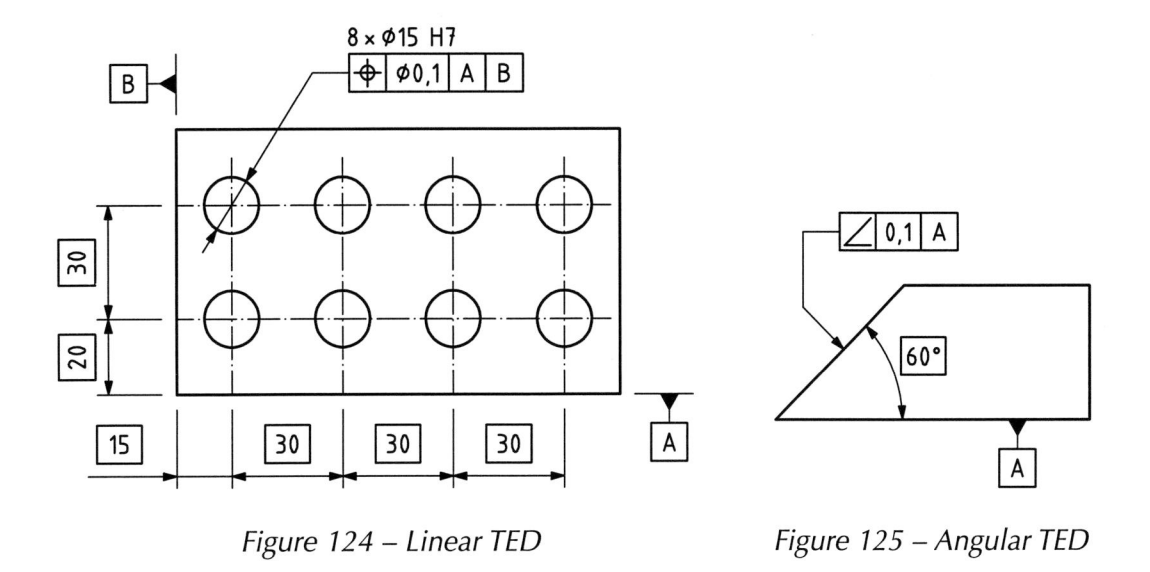

Figure 124 – Linear TED

Figure 125 – Angular TED

2.9.7 Restrictive specifications

If a tolerance of the same characteristic is applied to a restricted length, lying anywhere within the total extent of the feature, the value of the restricted length shall be added after the tolerance value and separated from it by an oblique stroke, as shown in Figure 126a. If two or more tolerances of the same characteristic are to be indicated, they may be combined as shown in Figure 126b.

Figure 126

2.9.8 Projected tolerance zone

The symbol Ⓟ after the tolerance value in the second compartment of the tolerance frame indicates a projected tolerance, as shown in Figures 127a and b. In this case, the tolerance applies to an extended feature.

The projected tolerance length shall be indicated in order to clearly define the toleranced feature.

The limits of the relevant portion of this extended feature shall be clearly defined and shall be indicated either directly or indirectly, as follows:

When indicating the projected tolerance length directly on a virtual integral feature representing the portion of the extended feature to be considered, this virtual feature shall be indicated by use of a long-dashed double-dotted narrow line, and the length of the extension shall be dimensioned with a theoretically exact dimension (TED) with the symbol Ⓟ prior to the value, as shown in Figure 127a.

When indicating the length of the projected toleranced feature indirectly in the tolerance frame, the value shall be specified after the symbol Ⓟ in the tolerance frame, as shown in Figure 127b. In this case the representation of the extended feature with a long-dashed double-dotted narrow line shall be omitted, as shown in Figures 128a and 128b. This indirect method only applies to blind holes.

a) Length of the extension by a TED

b) Length of extension within tolerance frame

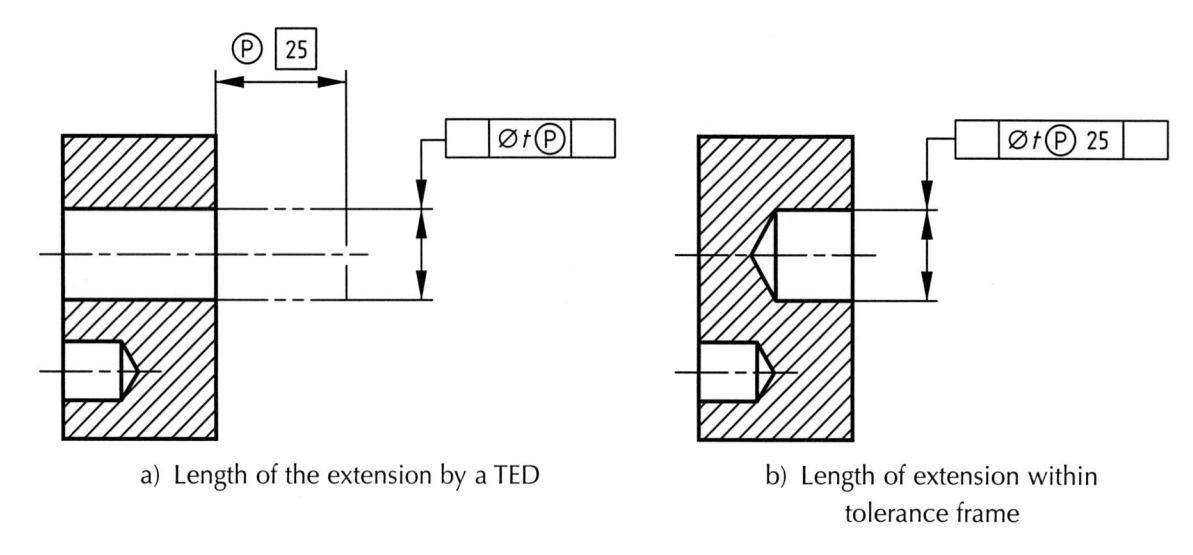

Figure 127

When indirectly indicating the length of the projected toleranced feature of a counter bored hole, the start of the projected zone is taken from the bottom of the counter bore, see Figure 128.

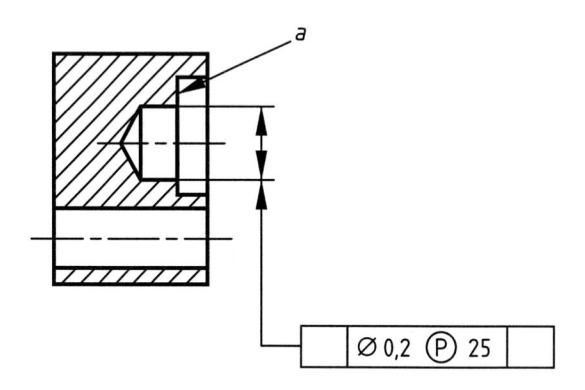

Figure 128

2.9.9 Maximum material requirement

The assembly of parts depends on the relationship between the actual size and the actual geometrical deviation of the features being fitted together, such as the bolt holes in two flanges and the bolts securing them.

The minimum assembly clearance occurs when each of the mating features is at its maximum material condition (MMC), e.g. largest bolt and smallest hole and when their geometrical deviations (e.g. positional deviation) are also at their maximum.

Assembly clearance increases to a maximum when the actual sizes of the assembled features are furthest from their maximum material conditions (e.g. smallest shaft and largest hole) and when the geometrical deviations (e.g. positional deviations) are zero. It follows that, if the actual sizes of a mating part do not reach their maximum material condition, the indicated geometrical tolerance may be increased without endangering the assembly of the other part.

This is called the 'maximum material principle' and is indicated on drawings by the symbol Ⓜ. The symbol is placed after the specified tolerance value, datum letter or both as appropriate, as shown in Figures 129, 130 and 131. (See BS EN ISO 2692 for detailed rules.)

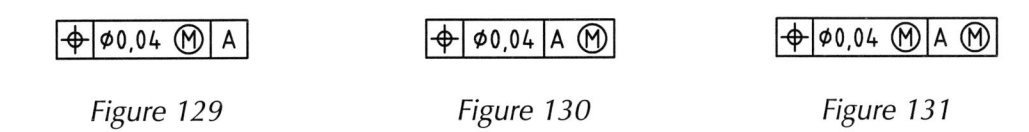

| ⊕ | ⌀0,04 Ⓜ | A |

Figure 129

| ⊕ | ⌀0,04 | A Ⓜ |

Figure 130

| ⊕ | ⌀0,04 Ⓜ | A Ⓜ |

Figure 131

The figures in this section are intended only as illustrations to aid the user in understanding the maximum material principle. In some instances, figures show added details for emphasis; in other instances, figures have deliberately been left incomplete. Numerical values of dimensions and tolerances have been given for illustrative purposes only.

For simplicity, the examples are limited to cylinders and planes.

Positional tolerance for a group of holes

The maximum material principle is most commonly used with positional tolerances, and therefore positional tolerancing has been used for the illustrations in this section.

NOTE: In the calculations of virtual size, it has been assumed that the pins and holes are at their maximum material condition and are of perfect form.

The indication on the drawing of the positional tolerance for a group of four holes is shown in Figure 132, and the interpretation in Figure 133.

The indication on the drawing of the positional tolerance for a group of four fixed pins that fit into the group of holes is shown in Figure 134.

- The minimum size of the holes is Ø8,1: this is the maximum material condition.
- The maximum size of the pins is Ø7,9: this is the maximum material condition.
- The difference between the maximum material condition of the holes and the pins is 8,1 − 7,9 = 0,2.

The sum of the positional tolerances for the holes and pins should not exceed this difference (0,2). In this example, this tolerance is equally distributed between holes and pins, i.e. the positional tolerance for the holes is Ø0,1, as shown in Figure 132 and the positional tolerance for the pins is also Ø0,1, as shown in Figure 134.

The tolerance zones of Ø0,1 are located at their theoretically exact positions, as shown in Figures 133 and 135. Depending on the actual size of each feature, the increase in the positional tolerance may be different for each feature.

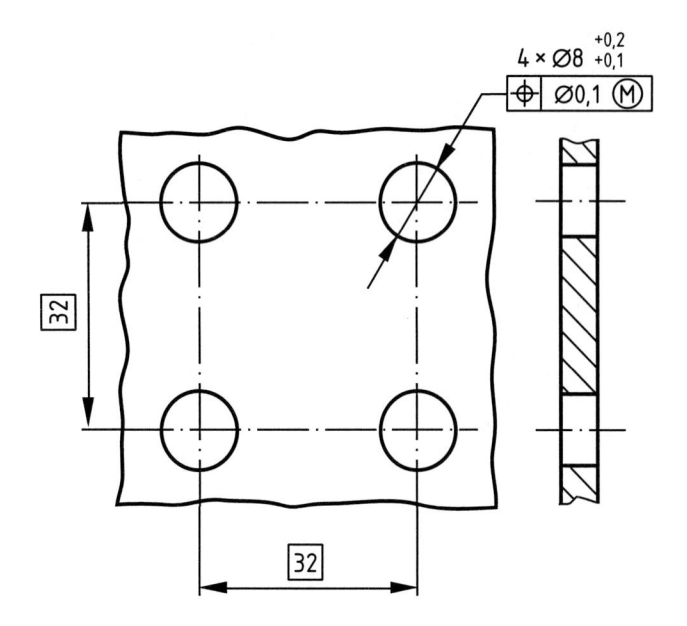

Figure 132 – Positional tolerance for a group of holes, indication on the drawing

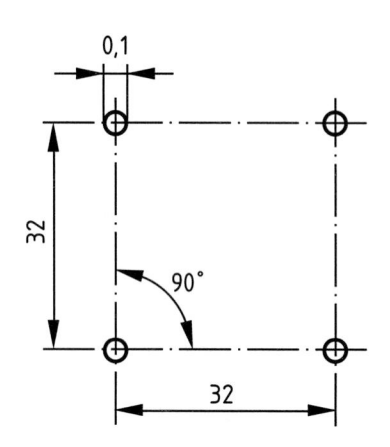

Figure 133 – Positional tolerance for a group of holes, interpretation

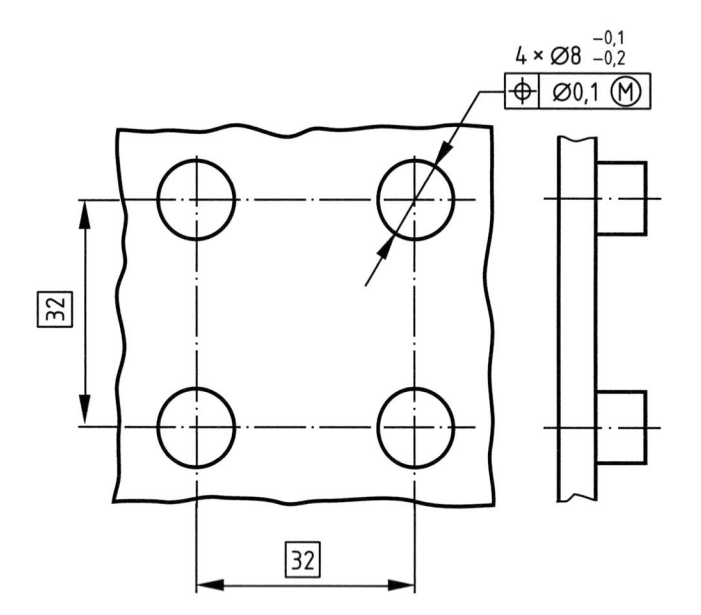

Figure 134 – Positional tolerance for a group of pins, indication on the drawing

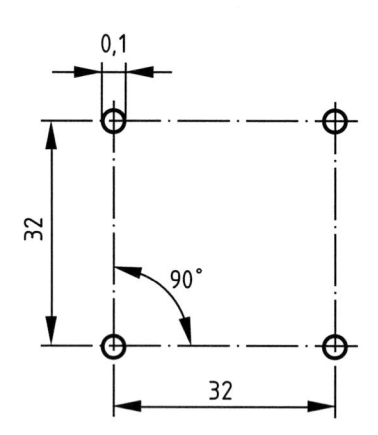

Figure 135 – Positional tolerance for a group of pins, interpretation

Figure 136 shows four cylindrical surfaces for each of the four holes all being at their maximum material condition and of perfect form. The axes are located at extreme positions within the tolerance zone. Figure 137 shows a larger scale version of Figure 136.

Figure 138 shows the corresponding pins at their maximum material condition. It can be seen from Figures 136–139 that assembly of the parts is still possible under the most unfavourable conditions.

The tolerance zone for the axis is Ø0,1. The maximum material condition of the hole is Ø8,1. All Ø8,1 circles, the axes of which are located at the extreme limit of the Ø0,1 tolerance zone, form an inscribed enveloping cylinder of Ø8. This cylinder is located at the theoretically exact position and forms the functional boundary for the surface of the hole.

One of the pins in Figure 138 is shown to a larger scale in Figure 139. The tolerance zone for the axis is Ø0,1. The maximum material condition of the pin is Ø7,9. All Ø7,9 circles, the axes of which are located at the extreme limit of the Ø0.1 tolerance zone, form a circumscribed enveloping cylinder of Ø8, which is the virtual condition of the pin.

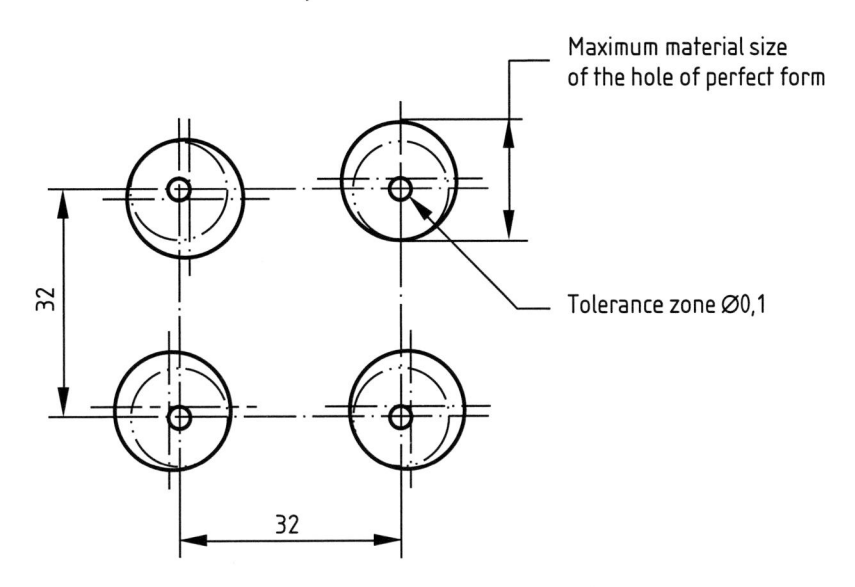

Figure 136 – Four holes (Figure 132) all at maximum material condition

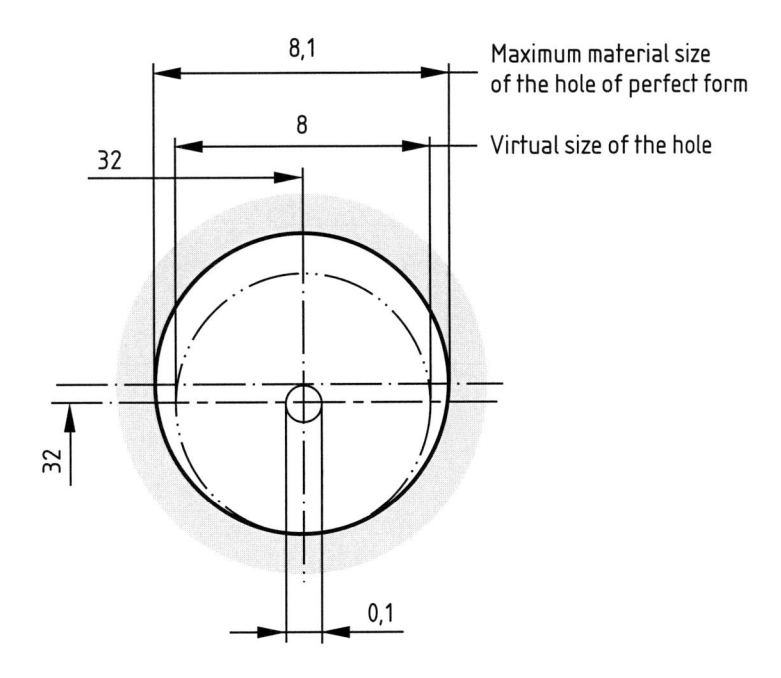

Figure 137 – Enlarged detail of Figure 136

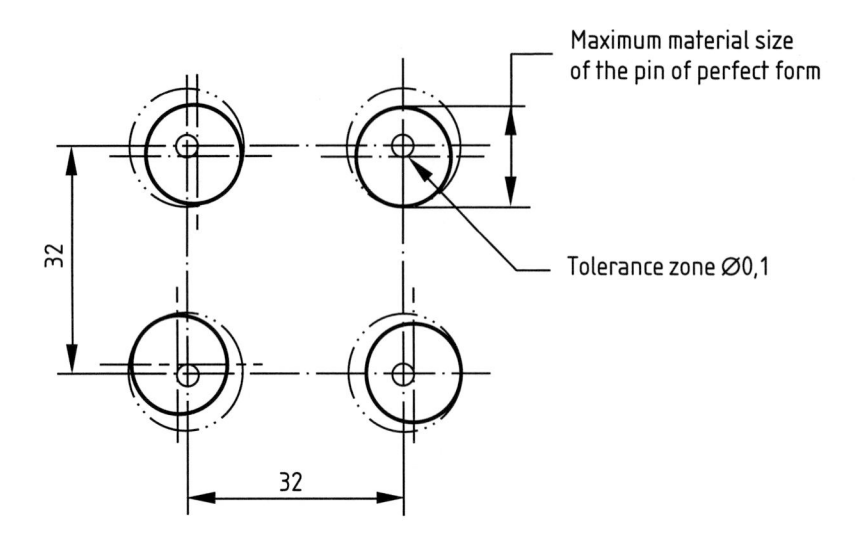

Figure 138 – Four pins (Figure 134) all at maximum material condition

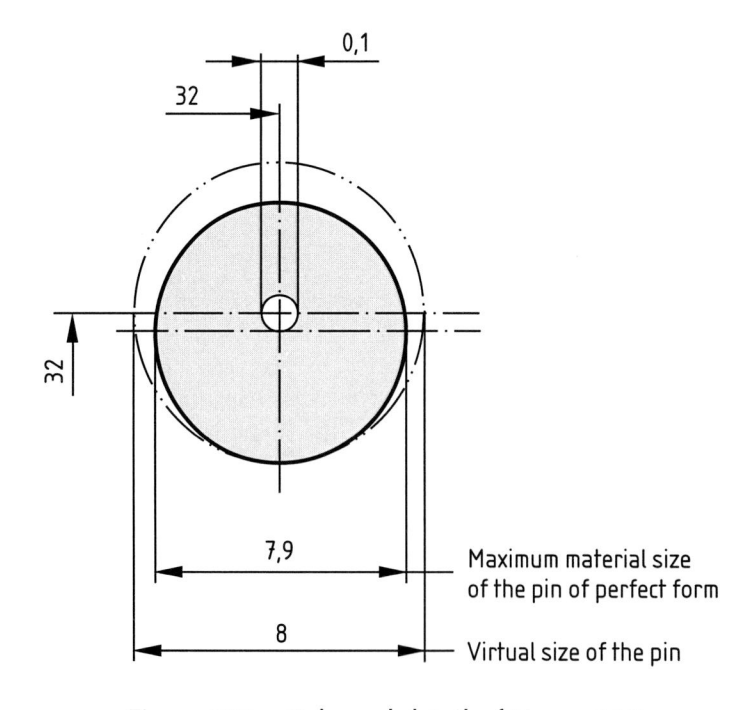

Figure 139 – Enlarged detail of Figure 138

When the size of the hole is larger than its maximum material condition and/or when the size of the pin is smaller than its maximum material condition, there is an increased clearance between the pin and hole, which can be used to increase the positional tolerances of the pin and/or the hole. Depending on the actual size of each feature, the increase in the positional tolerance may be different for each feature.

The extreme case is when the hole is at the least material condition, i.e. Ø8,2. Figure 140 shows that the axis of the hole may lie anywhere within a tolerance zone of Ø0,2 without the surface of the hole violating the cylinder of virtual size.

Figure 141 shows a similar situation with regard to the pins. When the pin is at the least material condition, i.e. Ø7,8, the diameter of the tolerance zone for position is Ø0,2.

Geometric tolerancing datums and datum systems

The increase in geometrical tolerance is applied to one part of the assembly without reference to the mating part. Assembly will always be possible even when the mating part is manufactured on the extreme limits of the tolerance in the direction most unfavourable for the assembly, because the combined deviation of size and geometry on neither part is exceeded, i.e. their virtual conditions are not violated.

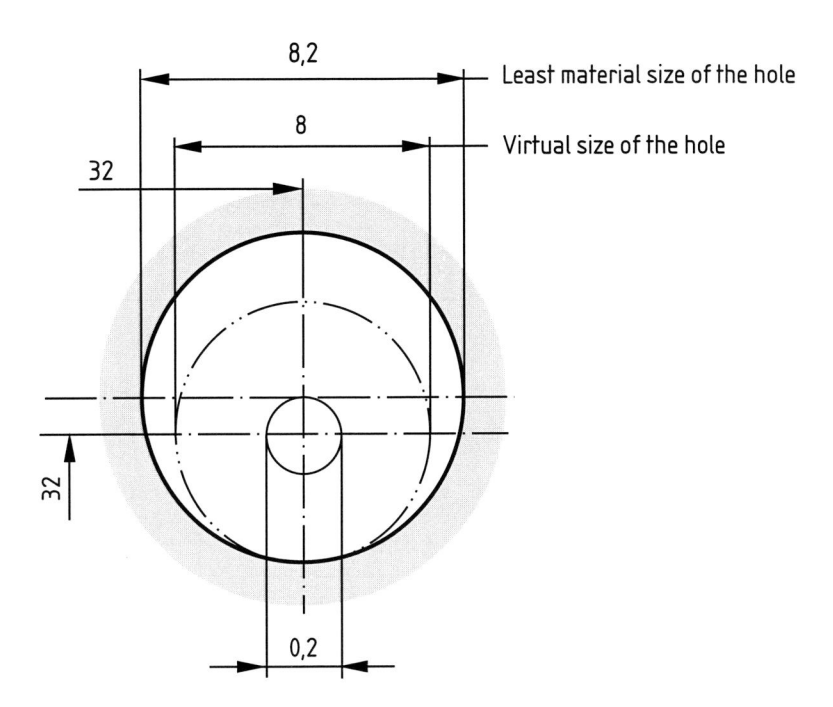

Figure 140 – Hole (Figure 129) at least material condition

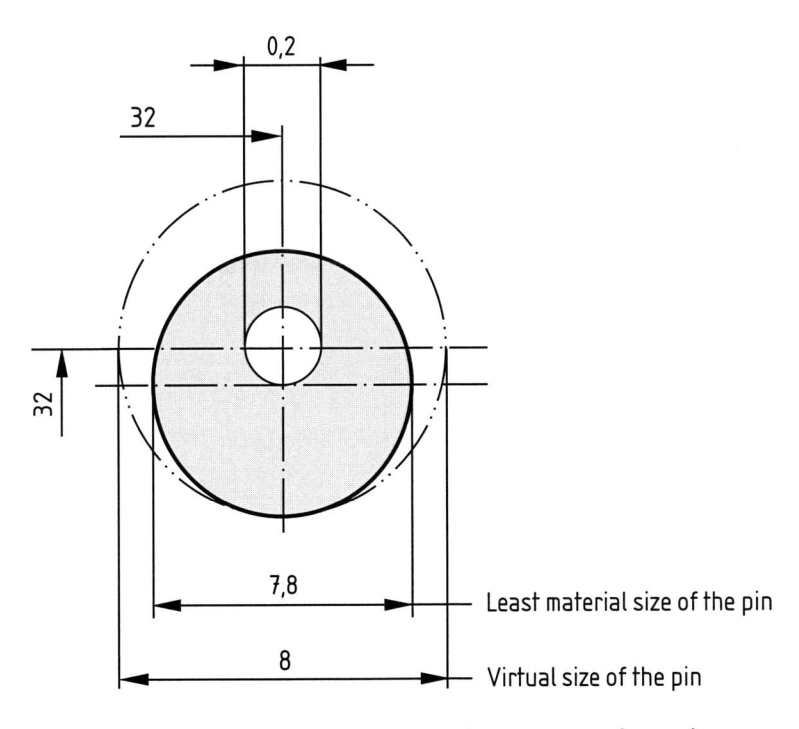

Figure 141 – Pin (Figure 131) at least material condition

2.9.10 Least material requirement

The least material requirement permits an increase in the stated geometrical tolerance when the concerned feature departs from its least material condition (LMC). This increase in the geometrical tolerance facilitates less scrapped components.

The least material requirement shall be indicated by the specification modifier symbol Ⓛ. The symbol shall be placed after the specified tolerance value, datum letter or both as appropriate, as shown in Figures 142, 143 and 144. It specifies:

- when applied to the toleranced feature, that the least material virtual condition (LMVC) should be fully contained within the material of the actual toleranced feature;
- when applied to the datum, that the boundary of perfect form at least material size may float within the material of the actual datum feature (without violating the actual datum feature surface).

See BS EN ISO 2692 for additional information.

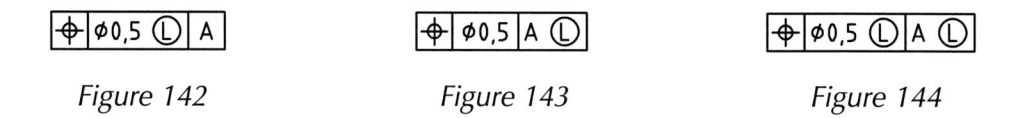

⌖ ⌀0,5 Ⓛ A	⌖ ⌀0,5 A Ⓛ	⌖ ⌀0,5 Ⓛ A Ⓛ
Figure 142	*Figure 143*	*Figure 144*

The least material requirement is illustrated in Figure 145. When the feature departs from its least material size, when it was at perfect form, an increase in positional tolerance is allowed, which is equal to the amount of such departure. An example of the application of least material requirement is illustrated in Figure 146.

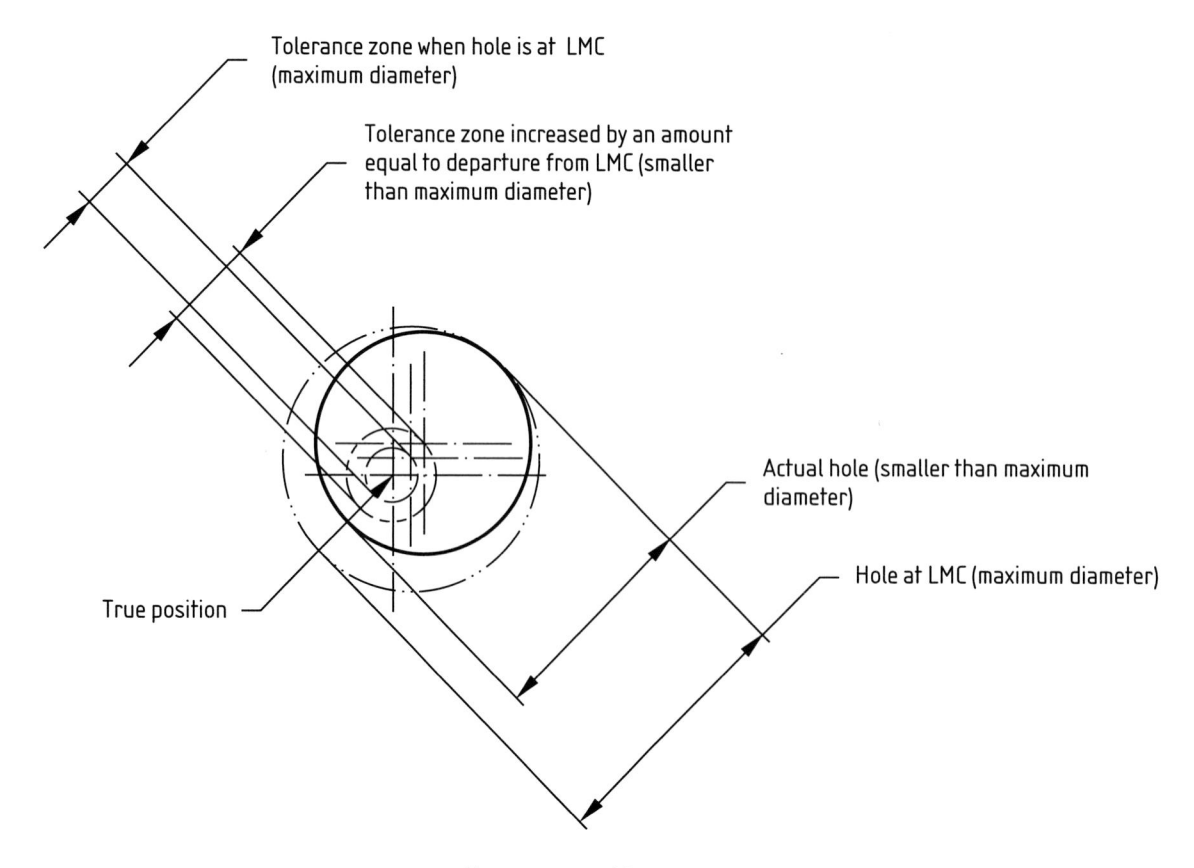

Figure 145 – Illustration of least material condition

Geometric tolerancing datums and datum systems

(a)

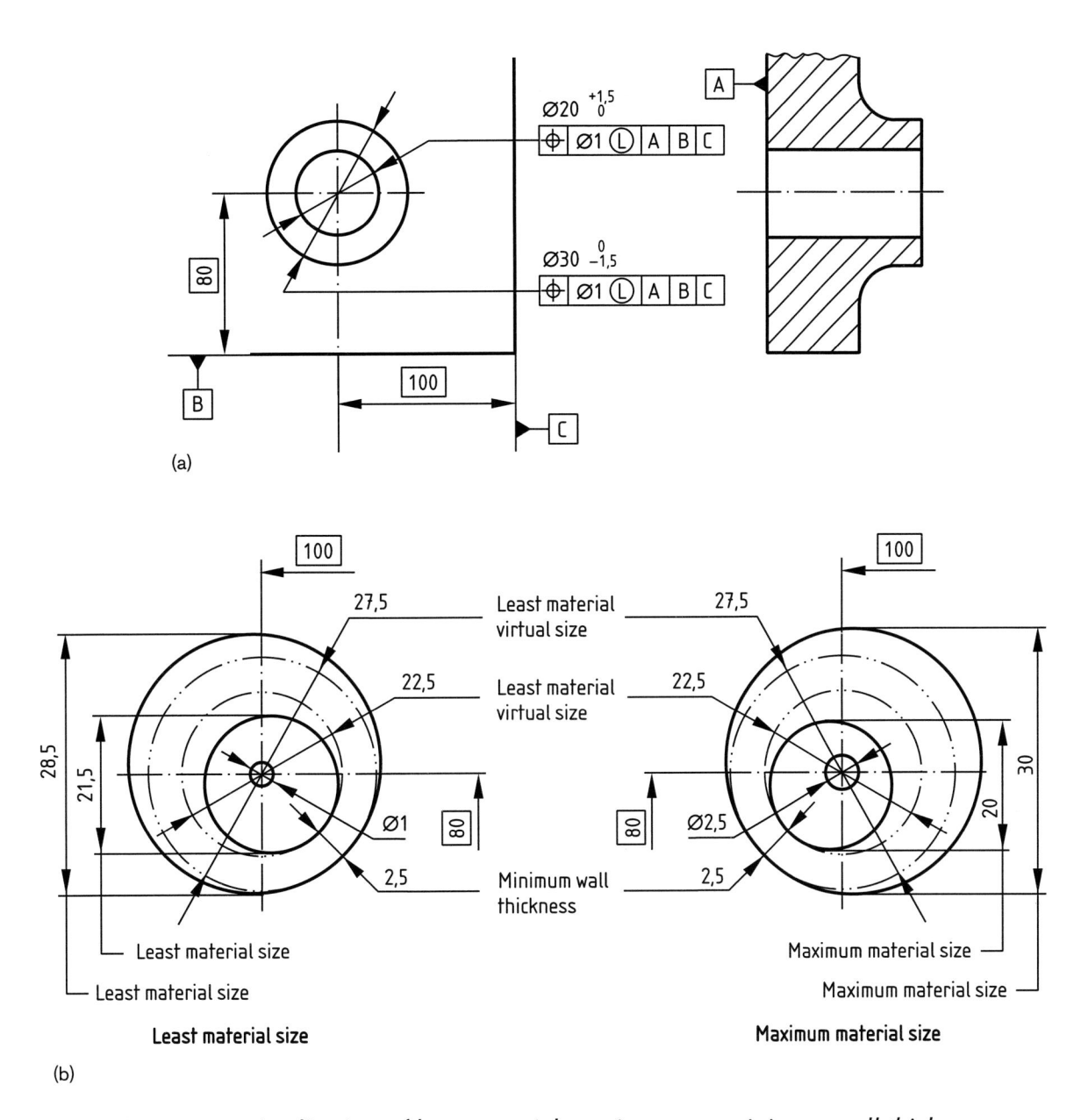

(b)

Figure 146 – Application of least material requirement – minimum wall thickness

2.9.11 Free state condition

The free state condition for non-rigid parts shall be indicated by the specification modifier symbol Ⓕ placed after the specified tolerance value, as shown in Figures 147 and 148. Application of the modifiers indicates that the tolerance applies only when the part is in its free state condition.

Several specification modifiers Ⓟ, Ⓜ, Ⓛ, Ⓕ and UZ may be used simultaneously in the same tolerance frame, as shown in Figure 149.

See BS ISO 10579 for additional information.

Figure 147

Figure 148

Figure 149

2.9.12 Intersection planes

Intersection planes can be used in 3D annotation to replicate the application of view dependent tolerances in 2D annotation, e.g. straightness of a line in a plane, profile of any line, orientation of a line element of a feature (LE), 'all around' specification for lines or surfaces. Intersection planes can also be used in 2D annotation if required.

Only surfaces belonging to one of the following shall be used to establish a family of intersection planes:

- revolute (e.g. a cone or a torus)
- cylindrical (i.e. a cylinder)
- planar (i.e. a plane)

The intersection plane is specified through an intersection plane indicator placed as an extension to the tolerance frame, as shown in Figures 152 to 155. The symbol defining how the intersection plane is derived from the datum is placed in the first compartment of the intersection plane indicator, as shown in Figure 150. The letter identifying the datum used to establish the intersection plane is placed in the second compartment of the intersection plane indicator, as shown in Figure 151.

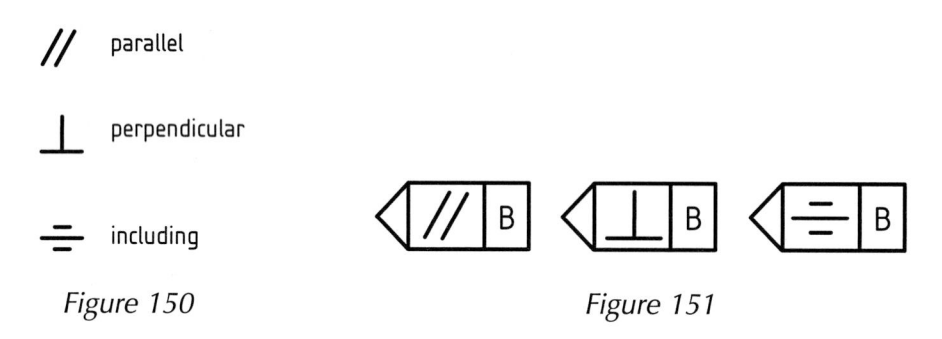

// parallel

⊥ perpendicular

= including

Figure 150

Figure 151

For geometrical specifications that include intersection plane indicators, the following applies:

- When the toleranced feature is a line on an integral feature, an intersection plane shall be indicated in 3D annotation to avoid misinterpretation of the toleranced feature, except in the case of straightness or circularity of a cylinder or a cone.
- The intersection plane is established parallel to, perpendicular to, or including the datum identified in the intersection plane indicator.
- The intersection plane is established parallel to, perpendicular to, or including the datum given in the intersection plane indicator without additional orientation constraints when the tolerance frame does not indicate datum(s).
- When the tolerance frame indicates datum(s), then the intersection plane is established parallel to, perpendicular to, or including the datum indicated in the intersection plane indicator with constraints (0°, 90° or an explicitly stated angle) from the datum(s) of the tolerance frame. The datums in the tolerance frame are applied in the specified order before the datum given in the intersection plane indicator is established.

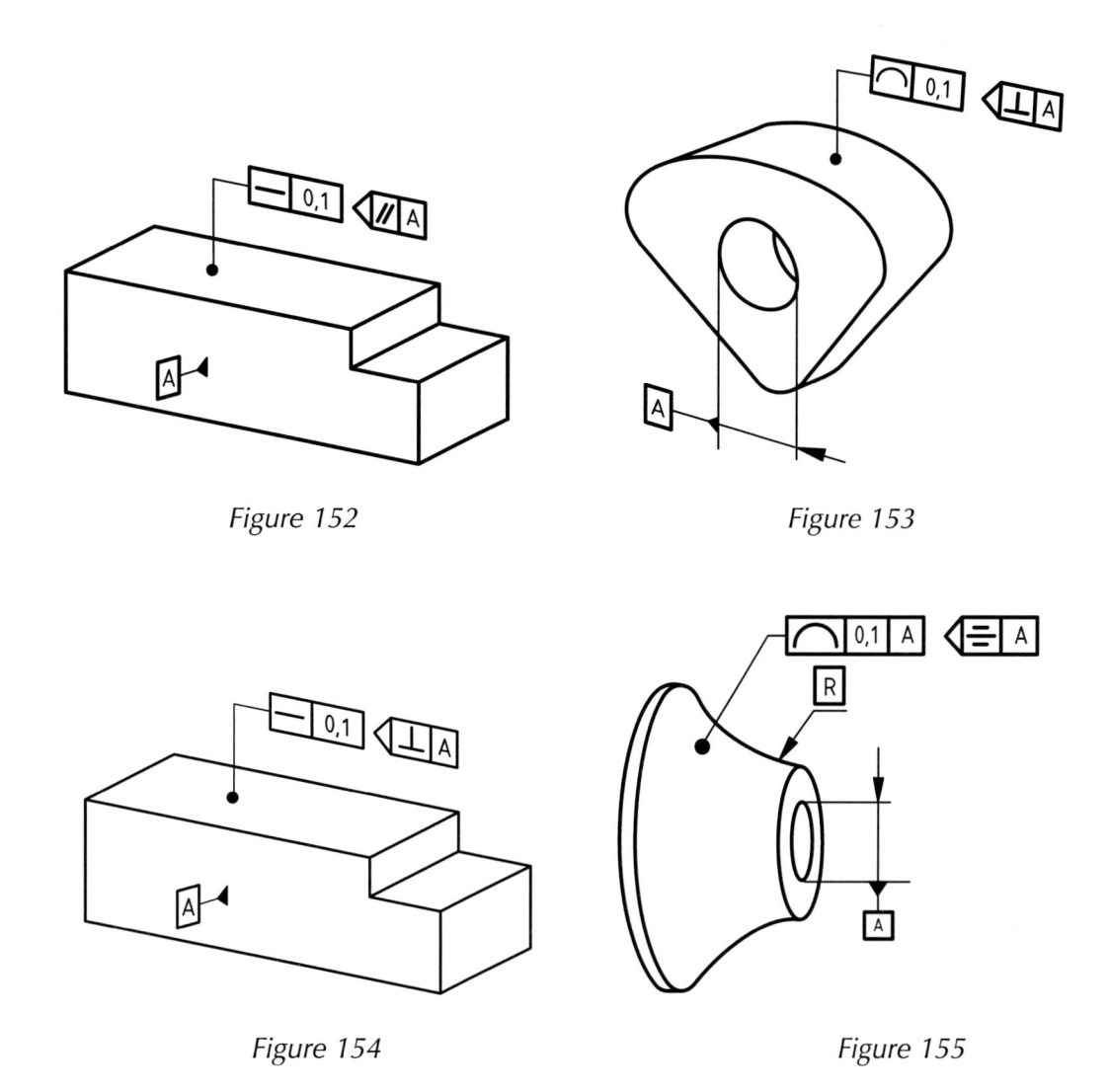

Geometric tolerancing datums and datum systems

Figure 152

Figure 153

Figure 154

Figure 155

A datum indicator and an intersection plane indicator located to the right of the tolerance frame shall be used to indicate an intersection plane, as shown in Table 3.

Table 3

Tolerance frame	Intersection plane indicator	Datum feature indicator
a)	b)	c)

The intersection plane indicator b) shall be placed to the right of the tolerance frame a). The intersection plane indicator shall indicate a datum letter in the second compartment. In the first compartment, a symbol is placed (parallel, perpendicular or symmetrical) and indicates how the intersection plane is related to the datum.

The datum corresponding to the datum letter allows building the intersection plane in accordance with the specified symbol.

The datum is defined from the datum feature identified by the datum feature indicator c).

2.9.13 Orientation planes

Orientation planes may be used in 3D annotation when the width of the tolerance zone is not normal to the specified geometry. Orientation planes can also be used in 2D annotation if required.

Only surfaces belonging to one of the following shall be used to establish orientation planes:

- revolute (e.g. a cone or a torus);
- cylindrical (i.e. a cylinder);
- planar (i.e. a plane).

The orientation plane is specified through an orientation plane indicator placed as an extension to the tolerance frame, as shown in Figures 158 and 159. The symbol defining how the orientation plane is derived from the datum is placed in the first compartment of the orientation plane indicator, as shown in Figures 156 and 157. The letter identifying the datum used to establish the orientation plane is placed in the second compartment of the orientation plane indicator.

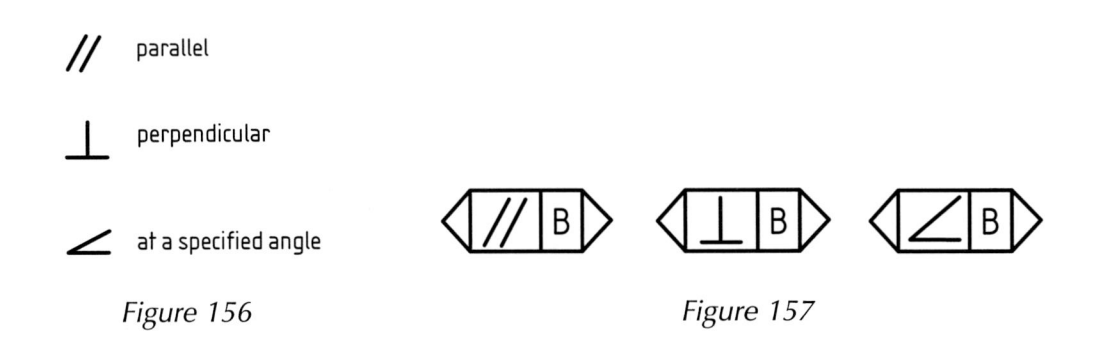

//   parallel

⊥   perpendicular

∠   at a specified angle

⟨ // | B ⟩   ⟨ ⊥ | B ⟩   ⟨ ∠ | B ⟩

Figure 156   *Figure 157*

An orientation plane shall be indicated in 3D annotation when:

- the width of the tolerance zone is not normal to the specified geometry; or,
- the toleranced feature is a point, or a median line toleranced in one Cartesian direction.

For geometrical specifications that include orientation plane indicators, the following applies.

- The orientation plane is established parallel to, perpendicular to, or at a defined angle from the datum indicated in the orientation plane indicator.
- The orientation plane is established without additional constraint of orientation when the tolerance frame does not indicate datum(s).
- When the tolerance frame indicates datum(s), then the orientation plane is established parallel to, perpendicular to, or at a defined angle from the datum indicated in the orientation plane indicator, with constraints (0°, 90° or an explicitly stated angle) from the datum(s) of the tolerance frame. The datums in the tolerance frame are applied in the specified order before the datum given in the orientation plane indicator is established.
- When the orientation plane is defined with an angle different from zero or 90°, the angularity symbol shall be used and an explicit theoretical angle (TED) shall be defined between the orientation plane and the datum of the orientation plane indicator.

Geometric tolerancing datums and datum systems

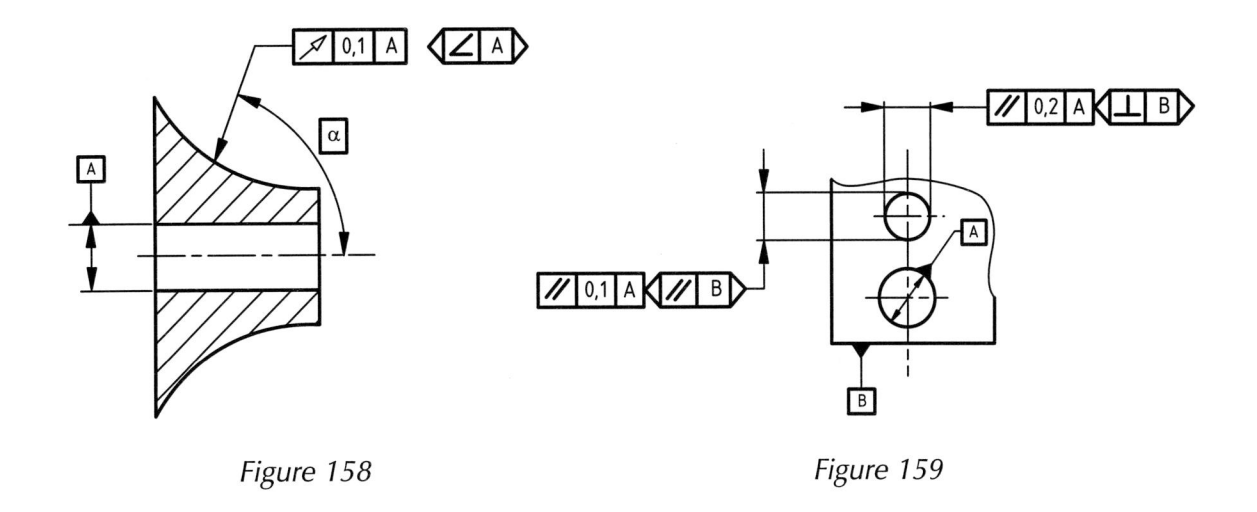

Figure 158

Figure 159

NOTE: The examples shown in Figures 156 and 157 are illustrated in 2D, in this application the orientation plane indicators are superfluous as the specification can be understood without them.

A datum indicator and an orientation plane indicator located to the right of the tolerance frame shall be used to indicate an orientation plane, as shown in Table 4.

Table 4

Datum indicator	Orientation plane indicator	Indication of geometrical specification using an orientation plane
[A]	⟨⊥|A⟩	◀—[|] ⟨⊥|A⟩
	The orientation plane indicator shall indicate the datum from which the orientation plane will be built, by a datum letter in its second compartment, and how the orientation of the width of the tolerance zone relates to the datum (parallel, perpendicular or inclined) by a symbol placed in its first compartment. In the case of an angle different from 0° or 90°, the angularity symbol shall be used and an explicit theoretical angle shall be defined between the orientation plane and the datum.	The orientation plane indicator shall be placed to the right of the tolerance frame.

2.10 Examples of geometrical tolerancing

Symbol	Definition of the tolerance zone	Indication and explanation in 2D	Indication and explanation in 3D
—	The tolerance zone, in the considered plane, is limited by two parallel straight lines a distance *t* apart and in the specified direction only. a Any distance. The tolerance zone is limited by two parallel planes a distance *t* apart.	***Straightness tolerance*** Any extracted (actual) line on the upper surface, parallel to the plane of projection in which the indication is shown, shall be contained between two parallel straight lines 0,1 apart. Any extracted (actual) generating line on the cylindrical surface shall be contained between two parallel planes 0,1 apart.	Any extracted (actual) line on the upper surface, parallel to the plane of projection in which the indication is shown, shall be contained between two parallel straight lines 0,1 apart. Any extracted (actual) generating line on the cylindrical surface shall be contained between two parallel planes 0,1 apart.

Geometric tolerancing datums and datum systems

Symbol	Definition of the tolerance zone	Indication and explanation in 2D	Indication and explanation in 3D
		Straightness tolerance (continued)	
—	The tolerance zone is limited by a cylinder of diameter t, if the tolerance value is preceded by the symbol Ø.	The extracted (actual) median line of the cylinder to which the tolerance applies shall be contained within a cylindrical zone of diameter 0,08.	The extracted (actual) median line of the cylinder to which the tolerance applies shall be contained within a cylindrical zone of diameter 0,08.
		Flatness tolerance	
▱	The tolerance zone is limited by two parallel planes a distance t apart.	The extracted (actual) surface shall be contained between two parallel planes 0,08 apart.	The extracted (actual) surface shall be contained between two parallel planes 0,08 apart.

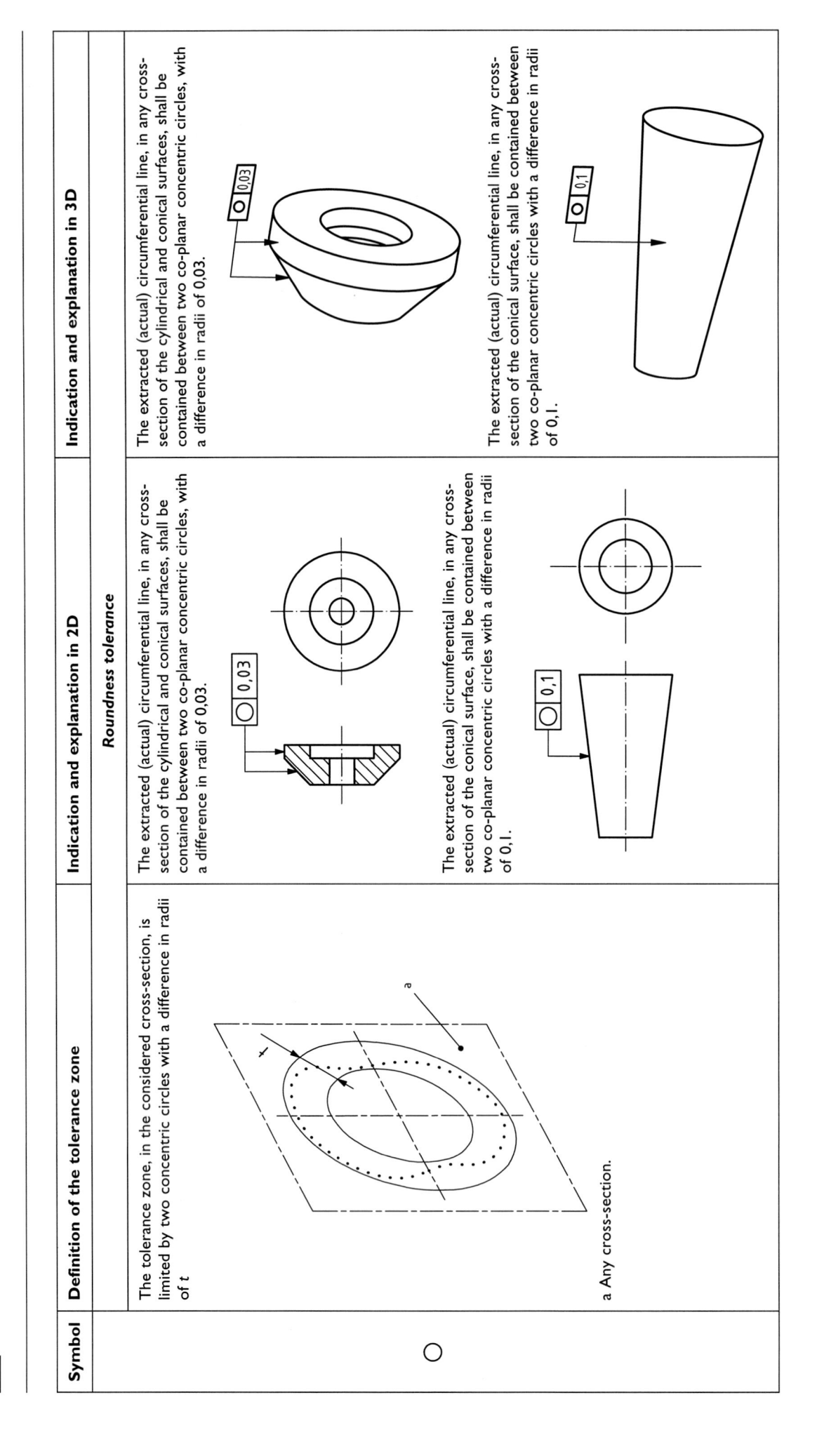

Symbol	Definition of the tolerance zone	Indication and explanation in 2D	Indication and explanation in 3D
	The tolerance zone, in the considered cross-section, is limited by two concentric circles with a difference in radii of t	**Roundness tolerance**	
○		The extracted (actual) circumferential line, in any cross-section of the cylindrical and conical surfaces, shall be contained between two co-planar concentric circles, with a difference in radii of 0,03.	The extracted (actual) circumferential line, in any cross-section of the cylindrical and conical surfaces, shall be contained between two co-planar concentric circles, with a difference in radii of 0,03.
		The extracted (actual) circumferential line, in any cross-section of the conical surface, shall be contained between two co-planar concentric circles with a difference in radii of 0,1.	The extracted (actual) circumferential line, in any cross-section of the conical surface, shall be contained between two co-planar concentric circles with a difference in radii of 0,1.

a Any cross-section.

Symbol	Definition of the tolerance zone	Indication and explanation in 2D	Indication and explanation in 3D
⌭	The tolerance zone is limited by two coaxial cylinders with a difference in radii of t.	**Cylindricity tolerance** The extracted (actual) cylindrical surface shall be contained between two coaxial cylinders with a difference in radii of 0,1.	The extracted (actual) cylindrical surface shall be contained between two coaxial cylinders with a difference in radii of 0,1

The Essential Guide to Technical Product Specification: Engineering Drawing

Symbol	Definition of the tolerance zone	Indication and explanation in 2D	Indication and explanation in 3D
	The tolerance zone is limited by two lines enveloping circles of diameter t, the centres of which are situated on a line having the theoretically exact geometrical form.	**Profile tolerance of a line not related to a datum** In each section, parallel to the plane of projection in which the indication is shown, the extracted (actual) profile line shall be contained between two equidistant lines enveloping circles of diameter 0,04, the centres of which are situated on a line having the theoretically exact geometrical form.	In each section, parallel to datum plane A, as specified by the intersection plane indicator, the extracted (actual) profile line shall be contained between two equidistant lines enveloping circles of diameter 0,04, the centres of which are situated on a line having the theoretically exact geometrical form.

a Any distance.
b Plane perpendicular to the drawing plane in 2D example.

Geometric tolerancing datums and datum systems

Profile tolerance of a line related to a datum system

Symbol	Definition of the tolerance zone	Indication and explanation in 2D	Indication and explanation in 3D
⌒	The tolerance zone is limited by two lines enveloping circles of diameter t, the centres of which are situated on a line having the theoretically exact geometrical form with respect to datum plane A and datum plane B. a Datum A. b Datum B. c Plane parallel to datum A.	In each section, parallel to the plane of projection in which the indication is shown, the extracted (actual) profile line shall be contained between two equidistant lines enveloping circles of diameter 0,04, the centres of which are situated on a line having the theoretically exact geometrical form with respect to datum plane A and datum plane B.	In each section, parallel to datum plane A, as specified by the intersection plane indicator, the extracted (actual) profile line shall be contained between two equidistant lines enveloping circles of diameter 0,04, the centres of which are situated on a line having the theoretically exact geometrical form with respect to datum plane A and datum plane B.

Symbol	Definition of the tolerance zone	Indication and explanation in 2D	Indication and explanation in 3D
		Profile tolerance of a surface not related to a datum	
⌓	The tolerance zone is limited by two surfaces enveloping spheres of diameter t, the centres of which are situated on a surface having the theoretically exact geometrical form.	The extracted (actual) surface shall be contained between two equidistant surfaces enveloping spheres of diameter 0,02, the centres of which are situated on a surface having the theoretically exact geometrical form.	The extracted (actual) surface shall be contained between two equidistant surfaces enveloping spheres of diameter 0,02, the centres of which are situated on a surface having the theoretically exact geometrical form.

Geometric tolerancing datums and datum systems

Profile tolerance of a surface related to a datum

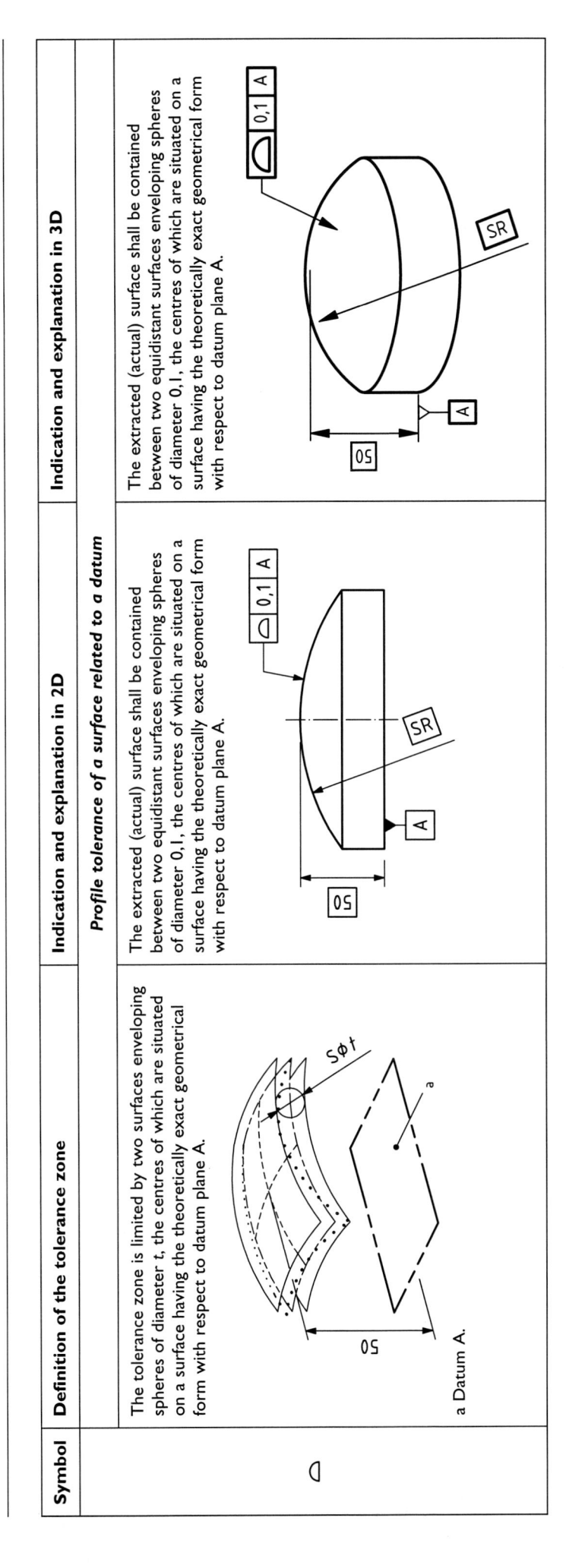

Symbol	Definition of the tolerance zone	Indication and explanation in 2D	Indication and explanation in 3D
⌓	The tolerance zone is limited by two surfaces enveloping spheres of diameter t, the centres of which are situated on a surface having the theoretically exact geometrical form with respect to datum plane A. a Datum A.	The extracted (actual) surface shall be contained between two equidistant surfaces enveloping spheres of diameter 0,1, the centres of which are situated on a surface having the theoretically exact geometrical form with respect to datum plane A.	The extracted (actual) surface shall be contained between two equidistant surfaces enveloping spheres of diameter 0,1, the centres of which are situated on a surface having the theoretically exact geometrical form with respect to datum plane A.

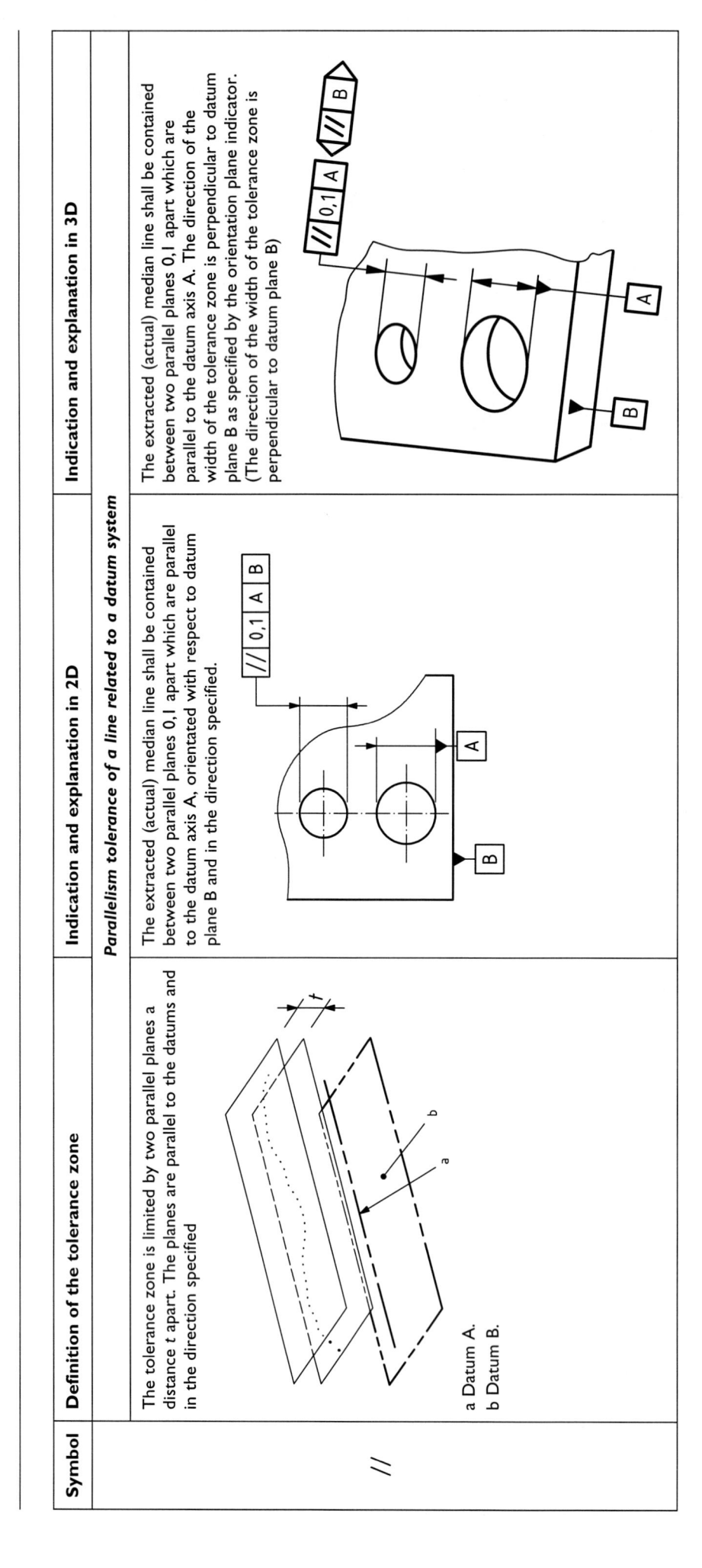

Symbol	Definition of the tolerance zone	Indication and explanation in 2D	Indication and explanation in 3D
		Parallelism tolerance of a line related to a datum system	
//	The tolerance zone is limited by two parallel planes a distance t apart. The planes are parallel to the datums and in the direction specified a Datum A. b Datum B.	The extracted (actual) median line shall be contained between two parallel planes 0,1 apart which are parallel to the datum axis A, orientated with respect to datum plane B and in the direction specified.	The extracted (actual) median line shall be contained between two parallel planes 0,1 apart which are parallel to the datum axis A. The direction of the width of the tolerance zone is perpendicular to datum plane B as specified by the orientation plane indicator. (The direction of the width of the tolerance zone is perpendicular to datum plane B)

Geometric tolerancing datums and datum systems

Symbol	Definition of the tolerance zone	Indication and explanation in 2D	Indication and explanation in 3D
		Parallelism tolerance of a line related to a datum system (continued)	
//			The extracted (actual) median line shall be contained between two parallel planes 0,1 apart, which are parallel to the datum axis A. The direction of the width of the tolerance zone is parallel to datum plane B as specified by the orientation plane indicator. (The direction of the width of the tolerance zone is parallel to datum plane B).

Symbol	Definition of the tolerance zone	Indication and explanation in 2D	Indication and explanation in 3D
		Parallelism tolerance of a line related to a datum system (continued)	
//			The extracted (actual) median line shall be contained between two pairs of parallel planes, parallel to the datum axis A, 0.1 and 0.2 apart respectively. The direction of the width of the tolerance zones is specified with respect to datum plane B by the orientation plane indicators.

Geometric tolerancing datums and datum systems

Symbol	Definition of the tolerance zone	Indication and explanation in 2D	Indication and explanation in 3D
		Parallelism tolerance of a line related to a datum line	
//	The tolerance zone is limited by a cylinder of diameter t, parallel to the datum, if the tolerance value is preceded by the symbol Ø. a Datum A.	The extracted (actual) median line shall be within a cylindrical zone of diameter 0,03 parallel to the datum axis A.	The extracted (actual) median line shall be within a cylindrical zone of diameter 0,03 parallel to the datum axis A.

Symbol	Definition of the tolerance zone	Indication and explanation in 2D	Indication and explanation in 3D
		Parallelism tolerance of *a line related to a datum surface*	
//	The tolerance zone is limited by two parallel planes a distance t apart and parallel to the datum a Datum B.	The extracted (actual) median line shall be contained between two parallel planes 0,01 apart which are parallel to the datum plane B.	The extracted (actual) median line shall be contained between two parallel planes 0,01 apart which are parallel to the datum plane B.

Geometric tolerancing datums and datum systems

Symbol	Definition of the tolerance zone	Indication and explanation in 2D	Indication and explanation in 3D
		Parallelism tolerance of a line related to a datum system	
//	The tolerance zone is limited by two parallel lines a distance t apart and oriented parallel to datum plane A, the lines lying in a plane parallel to datum plane B

a Datum A.
b Datum B. | Each extracted (actual) line shall be contained between two parallel lines 0,02 apart parallel to datum A and lying in a plane parallel to datum B. | Each extracted (actual) line, parallel to datum plane B as specified by the intersection plane indicator, shall be contained between two parallel lines 0,02 apart which are parallel to datum plane A. |

Symbol	Definition of the tolerance zone	Indication and explanation in 2D	Indication and explanation in 3D
		Parallelism tolerance of a surface related to a datum line	
//	The tolerance zone is limited by two parallel planes a distance t apart and parallel to the datum. a Datum C.	The extracted (actual) surface shall be contained between two parallel planes 0,1 apart which are parallel to the datum axis C.	The extracted (actual) surface shall be contained between two parallel planes 0,1 apart which are parallel to the datum axis C.

Geometric tolerancing datums and datum systems

Symbol	Definition of the tolerance zone	Indication and explanation in 2D	Indication and explanation in 3D
		Parallelism tolerance of a surface related to a datum surface	
//	The tolerance zone is limited by two parallel planes a distance t apart and parallel to the datum plane. a Datum D.	The extracted (actual) surface shall be contained between two parallel planes 0,01 apart which are parallel to datum plane D.	The extracted (actual) surface shall be contained between two parallel planes 0,01 apart which are parallel to datum plane D.

Symbol	Definition of the tolerance zone	Indication and explanation in 2D	Indication and explanation in 3D
		Perpendicularity tolerance	
⊥	The tolerance zone is limited by two parallel planes a distance t apart and perpendicular to the datum. a Datum A.	The extracted (actual) median line shall be contained between two parallel planes 0,06 apart that are perpendicular to datum axis A.	The extracted (actual) median line shall be contained between two parallel planes 0,06 apart that are perpendicular to the datum axis A.

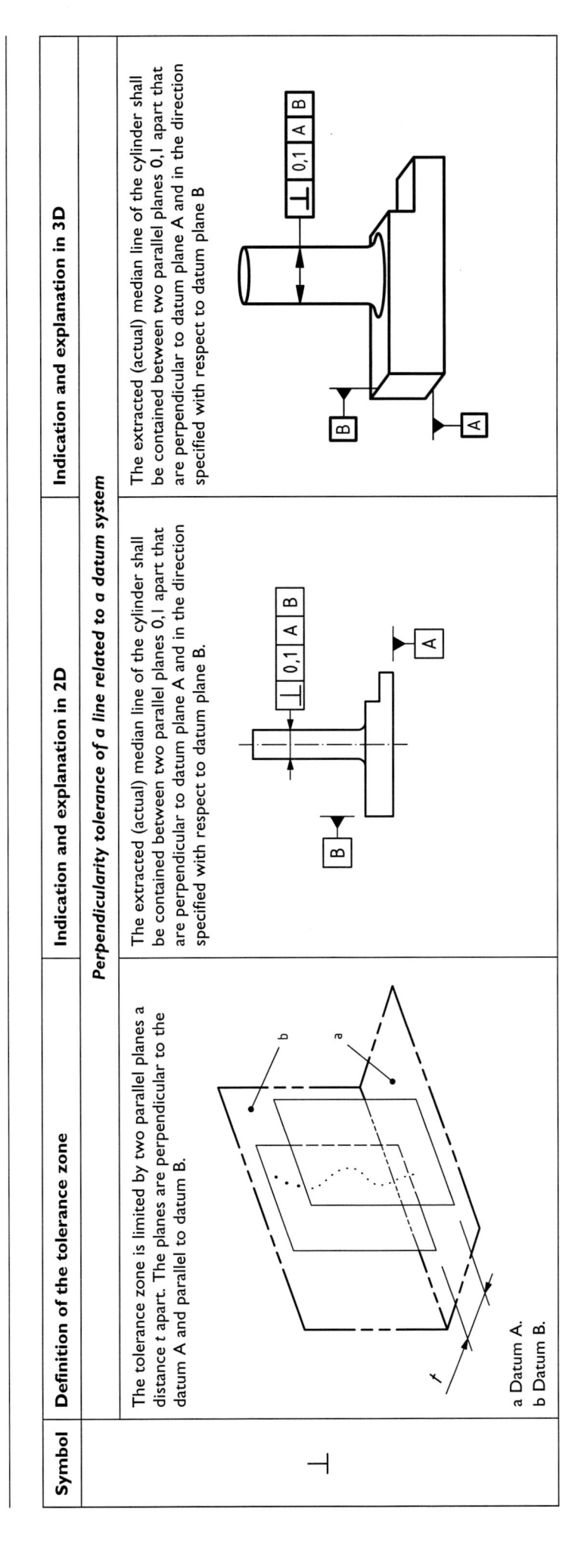

Geometric tolerancing datums and datum systems

Symbol	Definition of the tolerance zone	Indication and explanation in 2D	Indication and explanation in 3D
⊥	The tolerance zone is limited by two parallel planes a distance t apart. The planes are perpendicular to the datum A and parallel to datum B. a Datum A. b Datum B.	*Perpendicularity tolerance of a line related to a datum system* The extracted (actual) median line of the cylinder shall be contained between two parallel planes 0,1 apart that are perpendicular to datum plane A and in the direction specified with respect to datum plane B.	The extracted (actual) median line of the cylinder shall be contained between two parallel planes 0,1 apart that are perpendicular to datum plane A and in the direction specified with respect to datum plane B

Symbol	Definition of the tolerance zone	Indication and explanation in 2D	Indication and explanation in 3D
	Perpendicularity tolerance of a line related to a datum system (continued)		
⊥	The tolerance zone is limited by two pairs of parallel planes a distance 0,1 and 0,2 apart and perpendicular to each other. Both planes are perpendicular to the datum A, one pair of planes being parallel to datum B, the other pair being perpendicular to datum B (see below). a Datum A. b Datum B.	The extracted (actual) median line of the cylinder shall be contained between two pairs of parallel planes 0,1 and 0,2 apart, in the direction specified with respect to datum plane B, and perpendicular to each other. Both pairs of parallel planes shall be perpendicular to datum plane A.	The extracted (actual) median line of the cylinder shall be contained between two pairs of parallel planes, perpendicular to datum plane A, 0,1 and 0,2 apart, respectively. The direction of the width of the tolerance zones is specified with respect to datum plane B by the orientation plane indicators.

Symbol	Definition of the tolerance zone	Indication and explanation in 2D	Indication and explanation in 3D
⊥	*Perpendicularity tolerance of a line related to a datum system (continued)*		

0,2

a Datum A.
b Datum B.

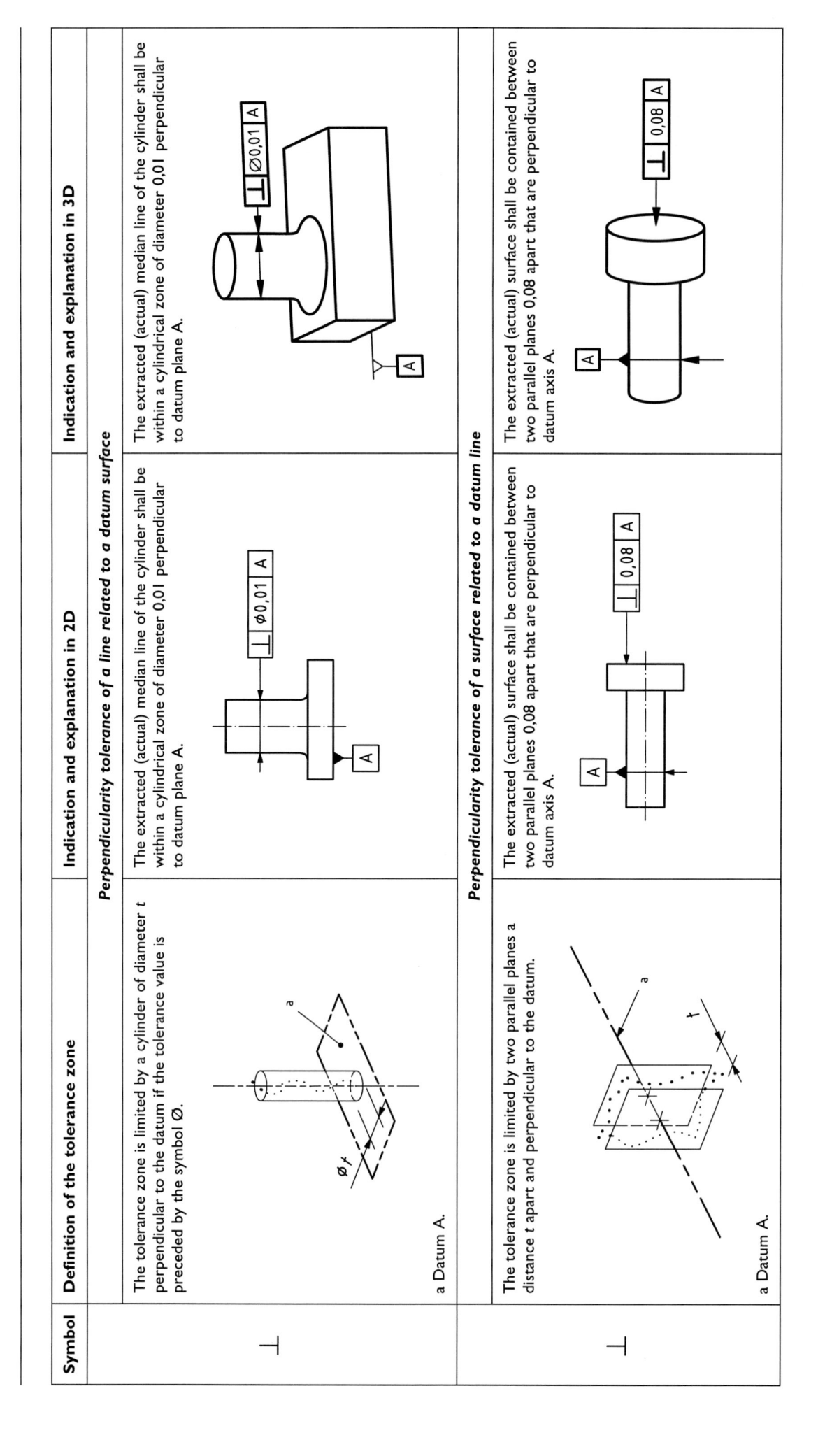

Symbol	Definition of the tolerance zone	Indication and explanation in 2D	Indication and explanation in 3D
		Perpendicularity tolerance of a line related to a datum surface	
⊥	The tolerance zone is limited by a cylinder of diameter t perpendicular to the datum if the tolerance value is preceded by the symbol ⌀. a Datum A.	The extracted (actual) median line of the cylinder shall be within a cylindrical zone of diameter 0,01 perpendicular to datum plane A.	The extracted (actual) median line of the cylinder shall be within a cylindrical zone of diameter 0,01 perpendicular to datum plane A.
		Perpendicularity tolerance of a surface related to a datum line	
⊥	The tolerance zone is limited by two parallel planes a distance t apart and perpendicular to the datum. a Datum A.	The extracted (actual) surface shall be contained between two parallel planes 0,08 apart that are perpendicular to datum axis A.	The extracted (actual) surface shall be contained between two parallel planes 0,08 apart that are perpendicular to datum axis A.

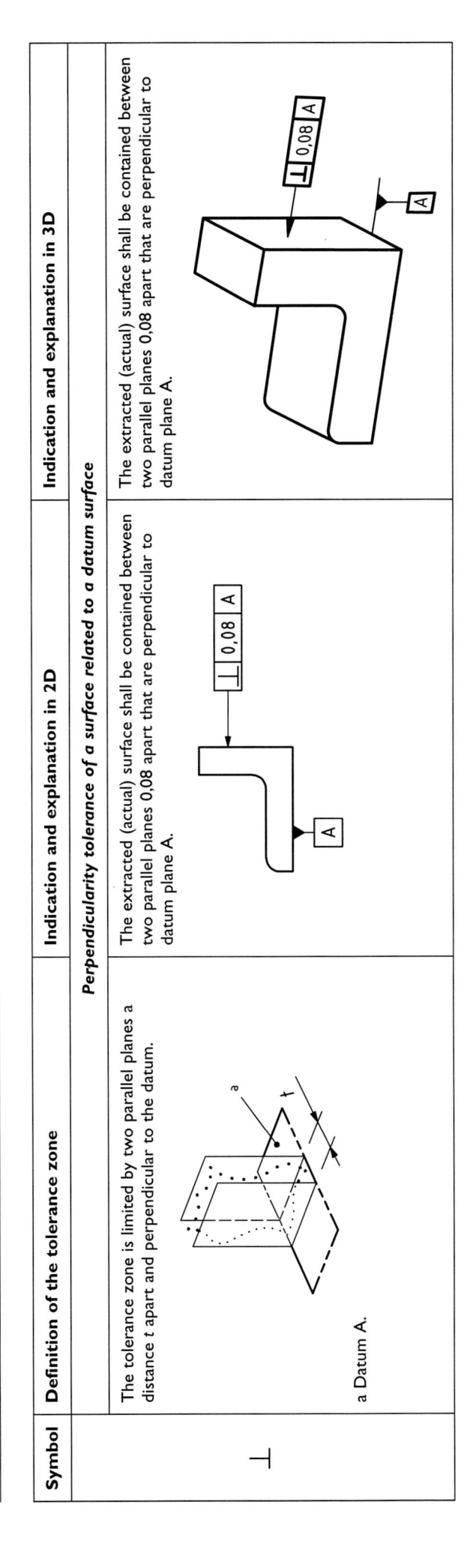

Geometric tolerancing datums and datum systems

Symbol	Definition of the tolerance zone	Indication and explanation in 2D	Indication and explanation in 3D
		Perpendicularity tolerance of a surface related to a datum surface	
⊥	The tolerance zone is limited by two parallel planes a distance t apart and perpendicular to the datum. a Datum A.	The extracted (actual) surface shall be contained between two parallel planes 0,08 apart that are perpendicular to datum plane A.	The extracted (actual) surface shall be contained between two parallel planes 0,08 apart that are perpendicular to datum plane A.

Symbol	Definition of the tolerance zone	Indication and explanation in 2D	Indication and explanation in 3D
		Angularity tolerance of a line related to a datum line	
∠	a) Line and datum line in the same plane: The tolerance zone is limited by two parallel planes a distance *t* apart and inclined at the specified angle to the datum. a Datum A-B.	The extracted (actual) median line shall be contained between two parallel planes 0,08 apart that are inclined at a theoretically exact angle of 60° to the common datum straight line A-B.	The extracted (actual) median line shall be contained between two parallel planes 0,08 apart that are inclined at a theoretically exact angle of 60° to the common datum straight line A-B.

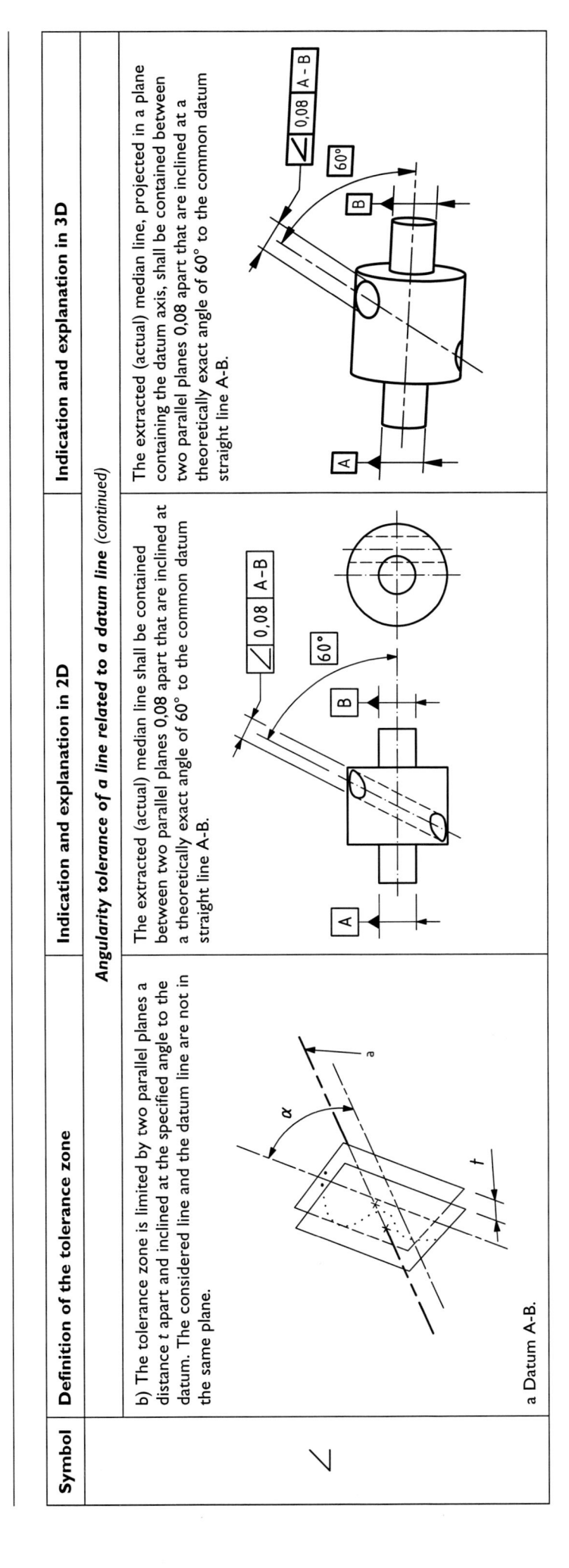

Geometric tolerancing datums and datum systems

Symbol	Definition of the tolerance zone	Indication and explanation in 2D	Indication and explanation in 3D
∠	b) The tolerance zone is limited by two parallel planes a distance t apart and inclined at the specified angle to the datum. The considered line and the datum line are not in the same plane. a Datum A-B.	*Angularity tolerance of a line related to a datum line (continued)* The extracted (actual) median line shall be contained between two parallel planes 0,08 apart that are inclined at a theoretically exact angle of 60° to the common datum straight line A-B.	The extracted (actual) median line, projected in a plane containing the datum axis, shall be contained between two parallel planes 0,08 apart that are inclined at a theoretically exact angle of 60° to the common datum straight line A-B.

Symbol	Definition of the tolerance zone	Indication and explanation in 2D	Indication and explanation in 3D
∠	The tolerance zone is limited by two parallel planes a distance t apart and inclined at the specified angle to the datum. a Datum A.	*Angularity tolerance of a line related to a datum surface* The extracted (actual) median line shall be contained between two parallel planes 0,08 apart that are inclined at a theoretically exact angle of 60° to datum plane A.	The extracted (actual) median line shall be contained between two parallel planes 0,08 apart that are inclined at a theoretically exact angle of 60° to the datum plane A.

Geometric tolerancing datums and datum systems

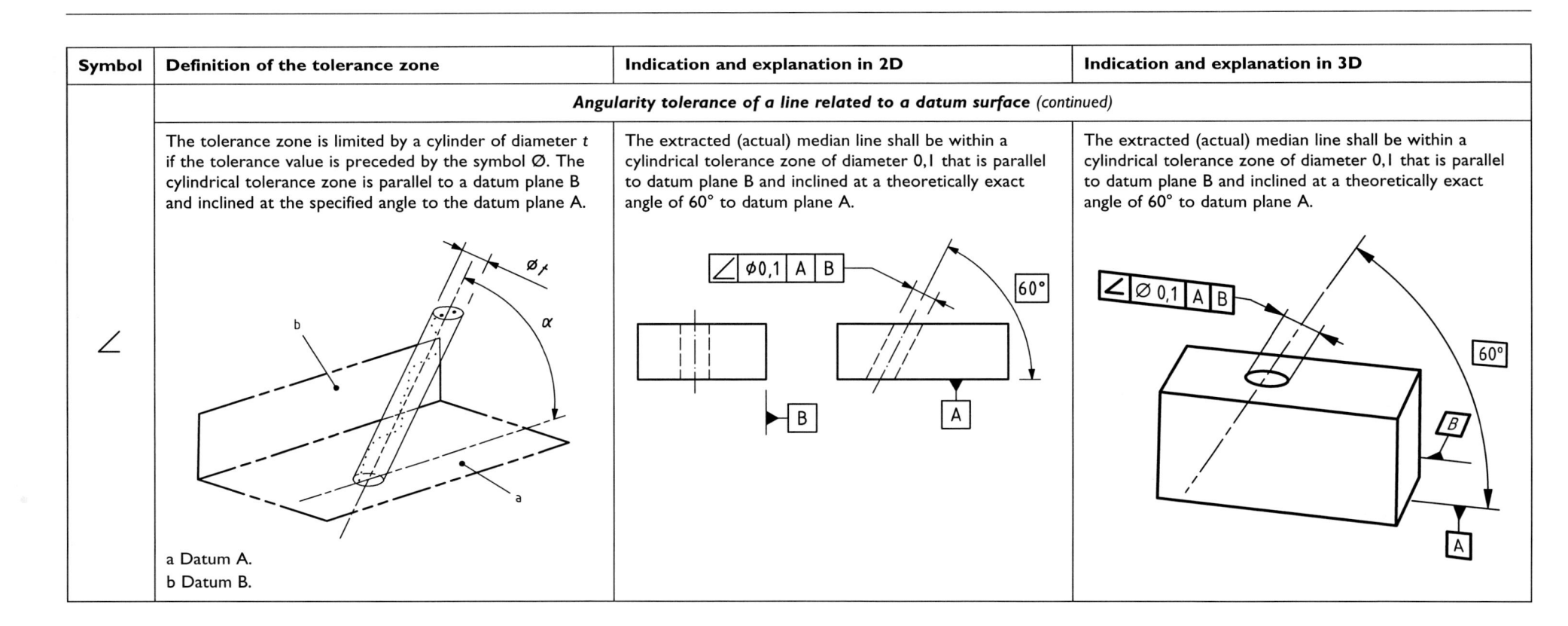

Symbol	Definition of the tolerance zone	Indication and explanation in 2D	Indication and explanation in 3D
		Angularity tolerance of a line related to a datum surface (continued)	
∠	The tolerance zone is limited by a cylinder of diameter t if the tolerance value is preceded by the symbol ⌀. The cylindrical tolerance zone is parallel to a datum plane B and inclined at the specified angle to the datum plane A. a Datum A. b Datum B.	The extracted (actual) median line shall be within a cylindrical tolerance zone of diameter 0,1 that is parallel to datum plane B and inclined at a theoretically exact angle of 60° to datum plane A.	The extracted (actual) median line shall be within a cylindrical tolerance zone of diameter 0,1 that is parallel to datum plane B and inclined at a theoretically exact angle of 60° to datum plane A.

Symbol	Definition of the tolerance zone	Indication and explanation in 2D	Indication and explanation in 3D
∠	The tolerance zone is limited by two parallel planes a distance t apart and inclined at the specified angle to the datum. a Datum A.	**Angularity tolerance of a surface related to a datum line** The extracted (actual) surface shall be contained between two parallel planes 0,1 apart that are inclined at a theoretically exact angle of 75° to datum axis A.	The extracted (actual) surface shall be contained between two parallel planes 0,1 apart that are inclined at a theoretically exact angle of 75° to datum axis A.

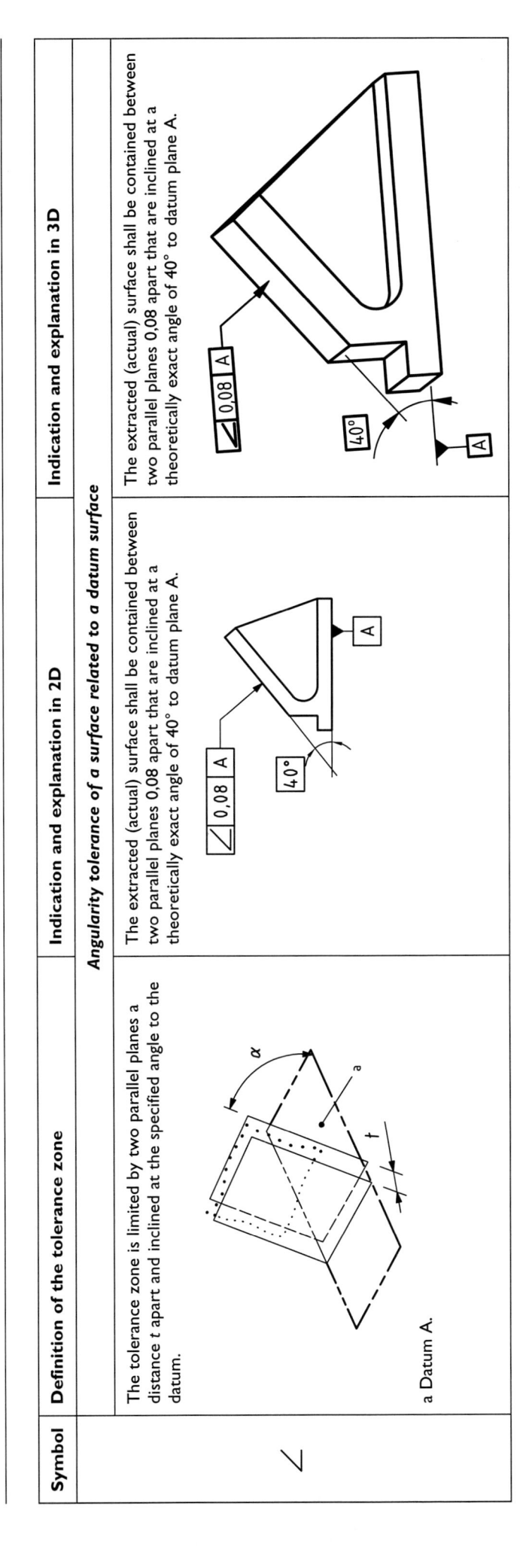

Geometric tolerancing datums and datum systems

Symbol	Definition of the tolerance zone	Indication and explanation in 2D	Indication and explanation in 3D
∠	The tolerance zone is limited by two parallel planes a distance t apart and inclined at the specified angle to the datum. a Datum A.	**Angularity tolerance of a surface related to a datum surface** The extracted (actual) surface shall be contained between two parallel planes 0,08 apart that are inclined at a theoretically exact angle of 40° to datum plane A.	The extracted (actual) surface shall be contained between two parallel planes 0,08 apart that are inclined at a theoretically exact angle of 40° to datum plane A.

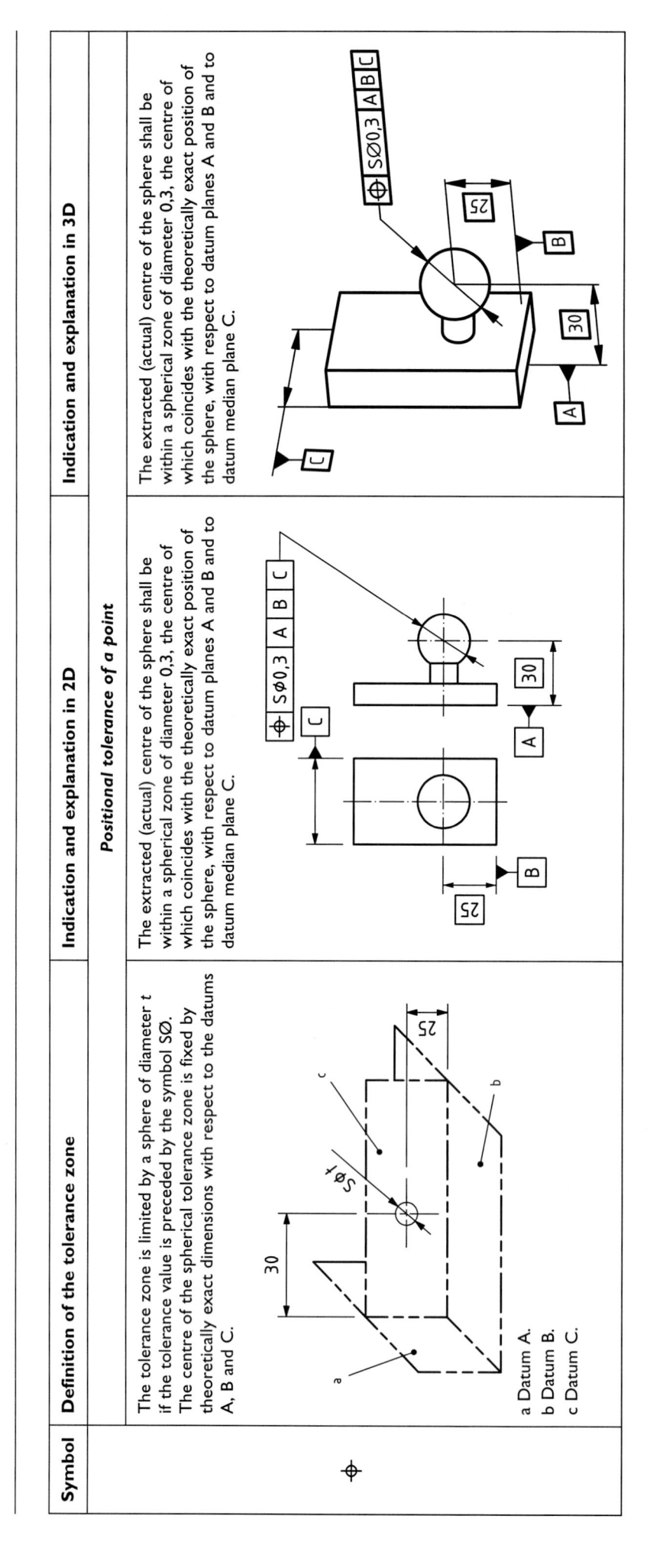

Symbol	Definition of the tolerance zone	Indication and explanation in 2D	Indication and explanation in 3D
		Positional tolerance of a point	
⌖	The tolerance zone is limited by a sphere of diameter t if the tolerance value is preceded by the symbol SØ. The centre of the spherical tolerance zone is fixed by theoretically exact dimensions with respect to the datums A, B and C. a Datum A. b Datum B. c Datum C.	The extracted (actual) centre of the sphere shall be within a spherical zone of diameter 0,3, the centre of which coincides with the theoretically exact position of the sphere, with respect to datum planes A and B and to datum median plane C.	The extracted (actual) centre of the sphere shall be within a spherical zone of diameter 0,3, the centre of which coincides with the theoretically exact position of the sphere, with respect to datum planes A and B and to datum median plane C.

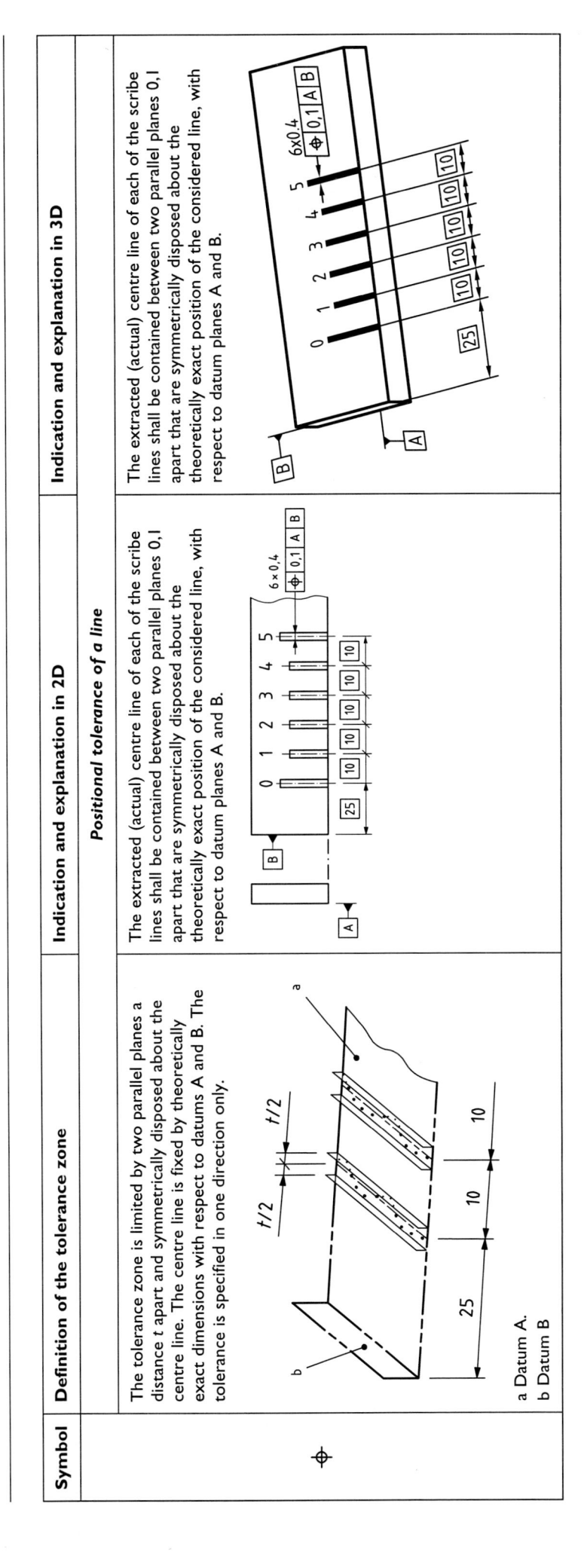

Geometric tolerancing datums and datum systems

Symbol	Definition of the tolerance zone	Indication and explanation in 2D	Indication and explanation in 3D
⌖	The tolerance zone is limited by two parallel planes a distance t apart and symmetrically disposed about the centre line. The centre line is fixed by theoretically exact dimensions with respect to datums A and B. The tolerance is specified in one direction only. a Datum A. b Datum B	**Positional tolerance of a line** The extracted (actual) centre line of each of the scribe lines shall be contained between two parallel planes 0,1 apart that are symmetrically disposed about the theoretically exact position of the considered line, with respect to datum planes A and B.	The extracted (actual) centre line of each of the scribe lines shall be contained between two parallel planes 0,1 apart that are symmetrically disposed about the theoretically exact position of the considered line, with respect to datum planes A and B.

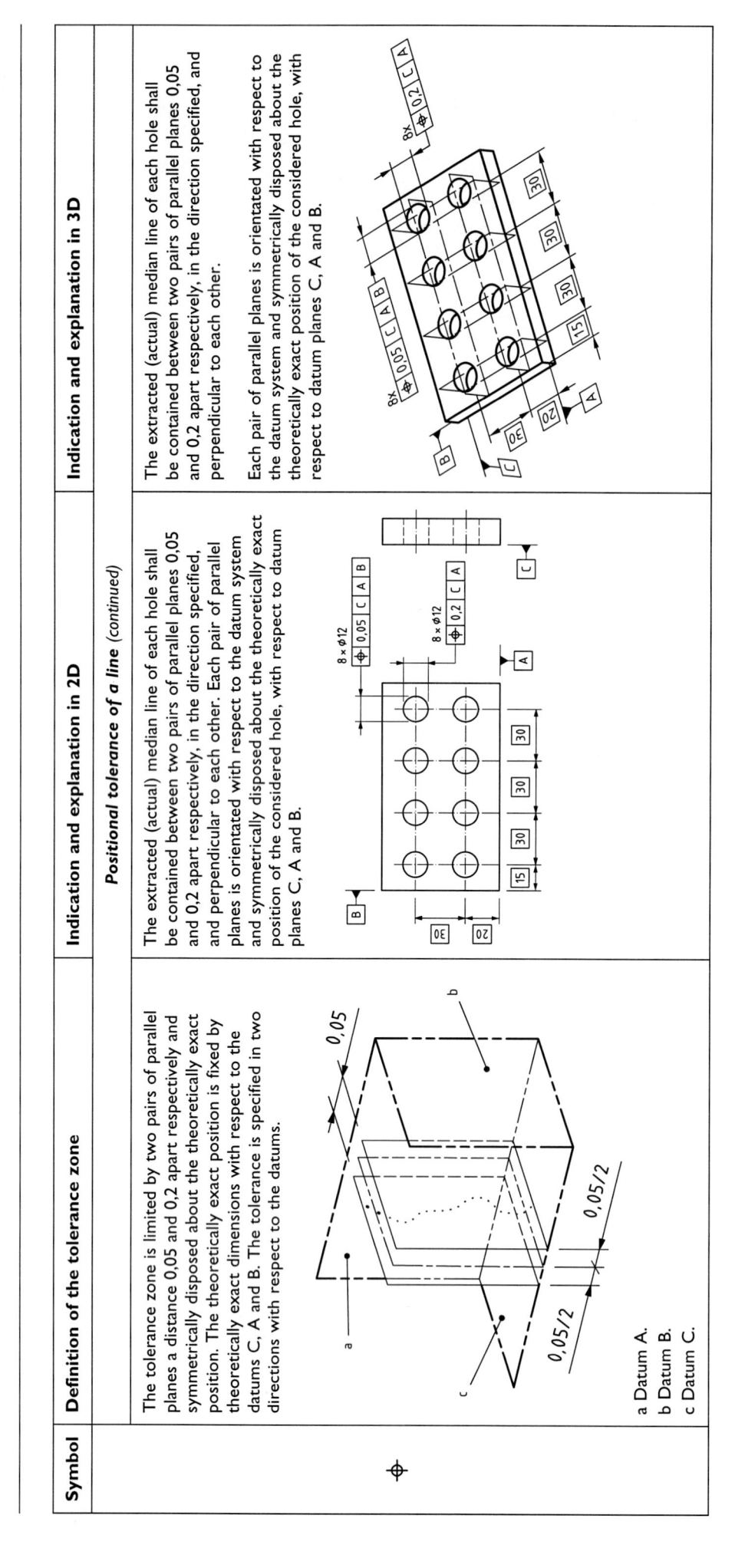

Symbol	Definition of the tolerance zone	Indication and explanation in 2D	Indication and explanation in 3D
		Positional tolerance of a line (continued)	
⌖	The tolerance zone is limited by two pairs of parallel planes a distance 0,05 and 0,2 apart respectively and symmetrically disposed about the theoretically exact position. The theoretically exact position is fixed by theoretically exact dimensions with respect to the datums C, A and B. The tolerance is specified in two directions with respect to the datums. a Datum A. b Datum B. c Datum C.	The extracted (actual) median line of each hole shall be contained between two pairs of parallel planes 0,05 and 0,2 apart respectively, in the direction specified, and perpendicular to each other. Each pair of parallel planes is orientated with respect to the datum system and symmetrically disposed about the theoretically exact position of the considered hole, with respect to datum planes C, A and B.	The extracted (actual) median line of each hole shall be contained between two pairs of parallel planes 0,05 and 0,2 apart respectively, in the direction specified, and perpendicular to each other. Each pair of parallel planes is orientated with respect to the datum system and symmetrically disposed about the theoretically exact position of the considered hole, with respect to datum planes C, A and B.

Geometric tolerancing datums and datum systems

95

Symbol	Definition of the tolerance zone	Indication and explanation in 2D	Indication and explanation in 3D
		Positional tolerance of a line (continued)	
⌖	[3D diagram showing tolerance zone with dimensions 0,2, 0,2/2, 0,2/2, with labels a, b, c] a Datum A. b Datum B. c Datum C.		

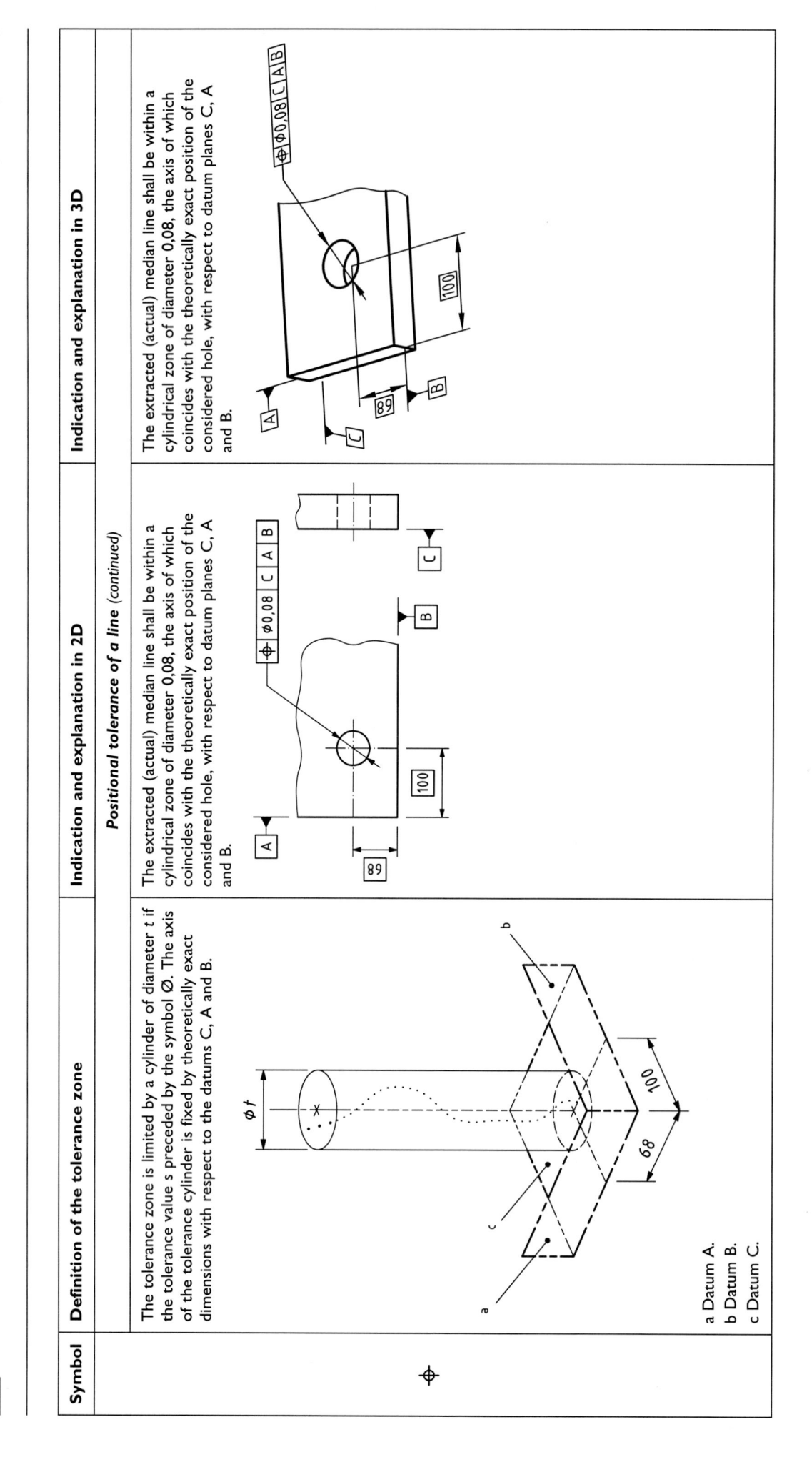

Symbol	Definition of the tolerance zone	Indication and explanation in 2D	Indication and explanation in 3D
		Positional tolerance of a line (continued)	
$\oplus$	The tolerance zone is limited by a cylinder of diameter t if the tolerance value s preceded by the symbol ⌀. The axis of the tolerance cylinder is fixed by theoretically exact dimensions with respect to the datums C, A and B. a Datum A. b Datum B. c Datum C.	The extracted (actual) median line shall be within a cylindrical zone of diameter 0,08, the axis of which coincides with the theoretically exact position of the considered hole, with respect to datum planes C, A and B.	The extracted (actual) median line shall be within a cylindrical zone of diameter 0,08, the axis of which coincides with the theoretically exact position of the considered hole, with respect to datum planes C, A and B.

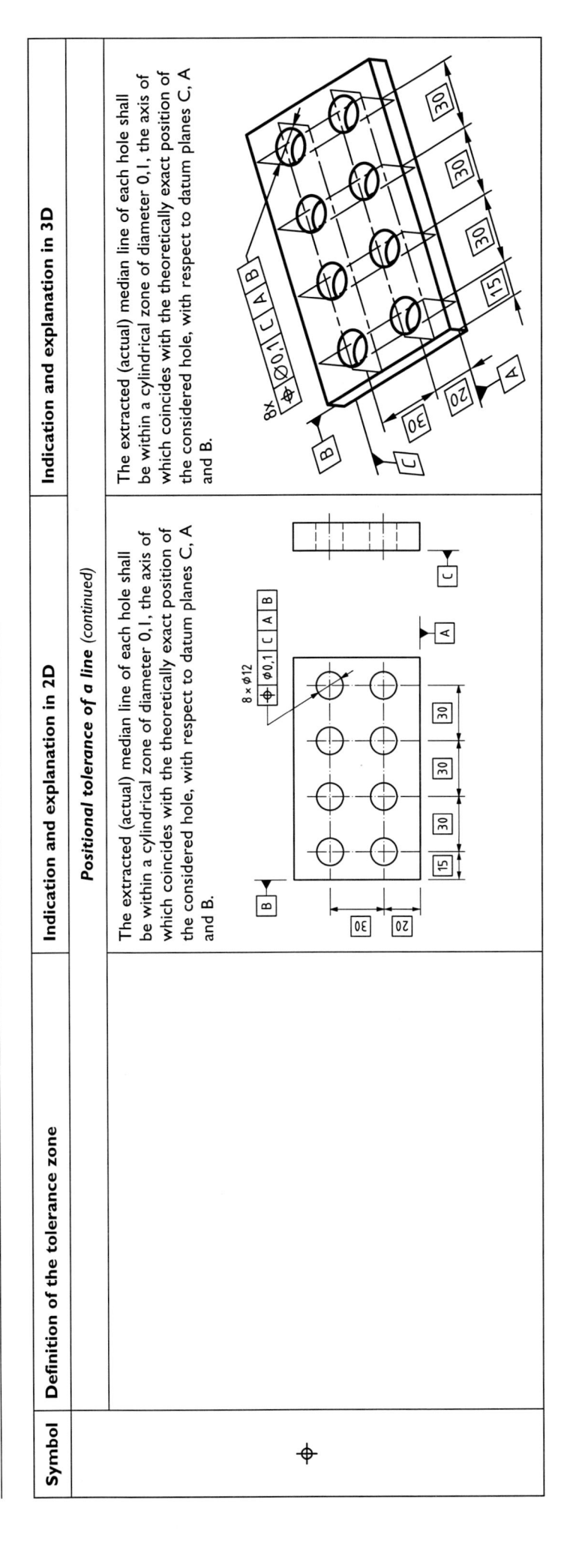

Geometric tolerancing datums and datum systems

Symbol	Definition of the tolerance zone	Indication and explanation in 2D	Indication and explanation in 3D
⌖		*Positional tolerance of a line* (continued)	
		The extracted (actual) median line of each hole shall be within a cylindrical zone of diameter 0,1, the axis of which coincides with the theoretically exact position of the considered hole, with respect to datum planes C, A and B.	The extracted (actual) median line of each hole shall be within a cylindrical zone of diameter 0,1, the axis of which coincides with the theoretically exact position of the considered hole, with respect to datum planes C, A and B.

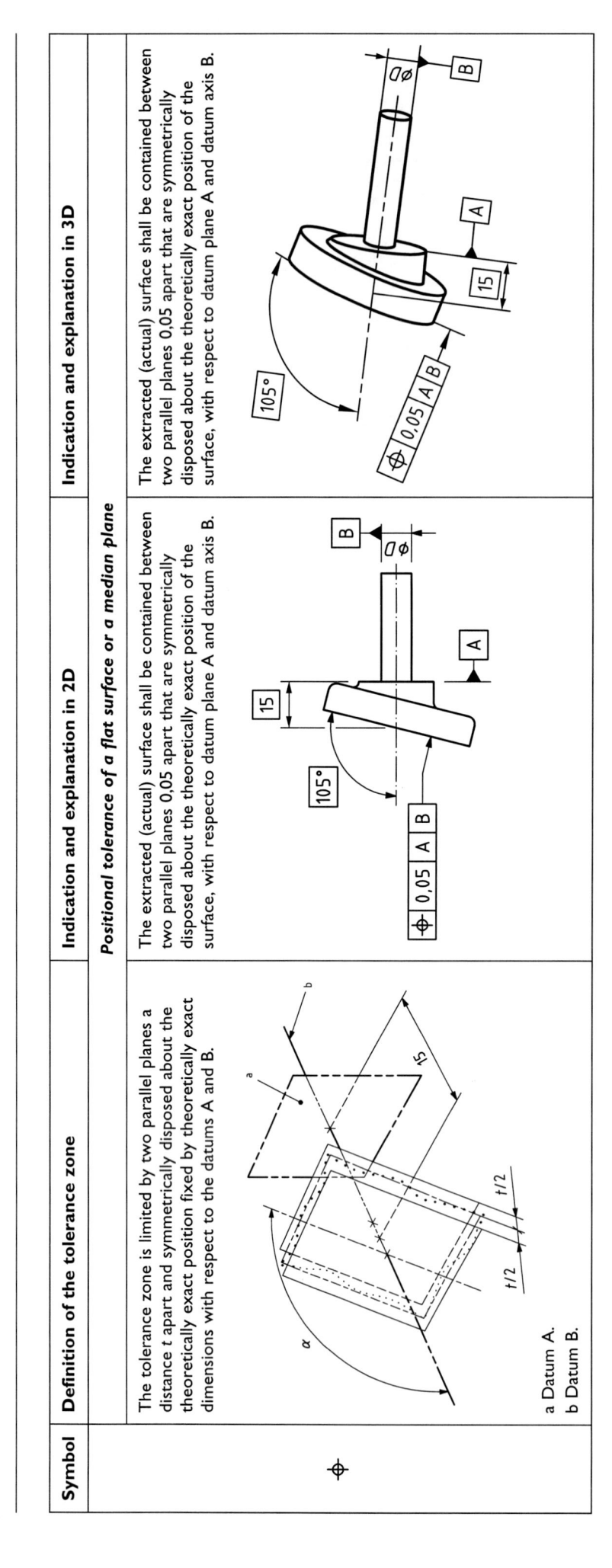

Symbol	Definition of the tolerance zone	Indication and explanation in 2D	Indication and explanation in 3D
		Positional tolerance of a flat surface or a median plane	
⊕	The tolerance zone is limited by two parallel planes a distance t apart and symmetrically disposed about the theoretically exact position fixed by theoretically exact dimensions with respect to the datums A and B. a Datum A. b Datum B.	The extracted (actual) surface shall be contained between two parallel planes 0,05 apart that are symmetrically disposed about the theoretically exact position of the surface, with respect to datum plane A and datum axis B.	The extracted (actual) surface shall be contained between two parallel planes 0,05 apart that are symmetrically disposed about the theoretically exact position of the surface, with respect to datum plane A and datum axis B.

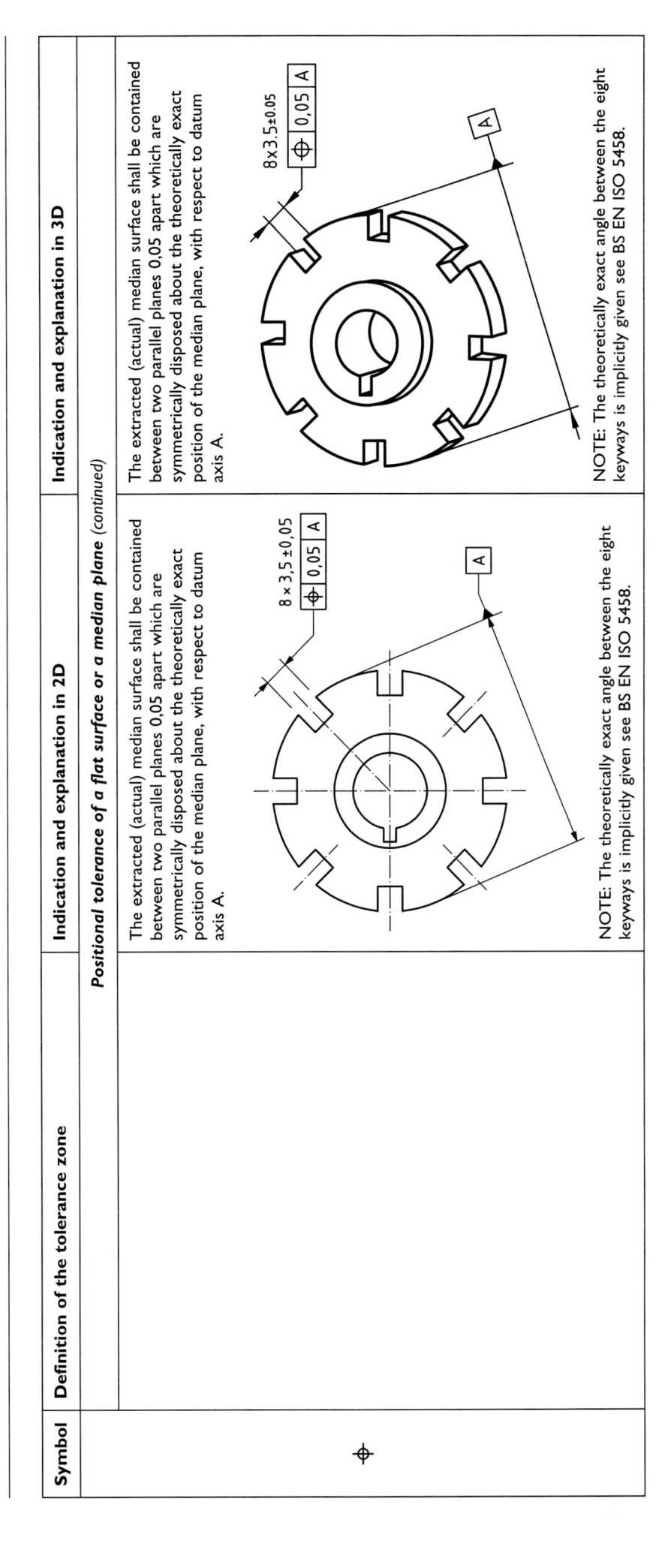

Geometric tolerancing datums and datum systems

Symbol	Definition of the tolerance zone	Indication and explanation in 2D	Indication and explanation in 3D
	Positional tolerance of a flat surface or a median plane (continued)		
⌖		The extracted (actual) median surface shall be contained between two parallel planes 0,05 apart which are symmetrically disposed about the theoretically exact position of the median plane, with respect to datum axis A. 8 × 3,5 ± 0,05 ⌖ 0,05 A NOTE: The theoretically exact angle between the eight keyways is implicitly given see BS EN ISO 5458.	The extracted (actual) median surface shall be contained between two parallel planes 0,05 apart which are symmetrically disposed about the theoretically exact position of the median plane, with respect to datum axis A. 8×3,5±0,05 ⌖ 0,05 A NOTE: The theoretically exact angle between the eight keyways is implicitly given see BS EN ISO 5458.

Symbol	Definition of the tolerance zone	Indication and explanation in 2D	Indication and explanation in 3D
		Concentricity tolerance of a point	
◎	The tolerance zone is limited by a circle of diameter t; the tolerance value shall be preceded by the symbol Ø. The centre of the circular tolerance zone coincides with the datum point. a Datum point A.	The extracted (actual) centre of the inner circle shall be within a circle of diameter 0,1 concentric with datum point A in the cross-section.	The extracted (actual) centre of the inner circle shall be within a circle of diameter 0,1 concentric with datum point A in the cross-section.

Geometric tolerancing datums and datum systems

Symbol	Definition of the tolerance zone	Indication and explanation in 2D	Indication and explanation in 3D
◎	The tolerance zone is limited by a cylinder of diameter t; the tolerance value shall be preceded by the symbol Ø. The axis of the cylindrical tolerance zone coincides with the datum. a Datum A-B.	**Coaxiality tolerance of an axis** The extracted (actual) median line of the toleranced cylinder shall be within a cylindrical zone of diameter 0,08 the axis of which is the common datum straight line A-B. ◎ Ø0,08 A-B The extracted (actual) median line of the toleranced cylinder shall be within a cylindrical zone of diameter 0,1 the median line of which is the datum axis A. ◎ Ø0,1 A	The extracted (actual) median line of the toleranced cylinder shall be within a cylindrical zone of diameter 0,08 the axis of which is the common datum straight line A-B. ◎ A Ø0,08 A-B The extracted (actual) median line of the toleranced cylinder shall be within a cylindrical zone of diameter 0,1 the median line of which is the datum axis A. ◎ Ø0,1 A

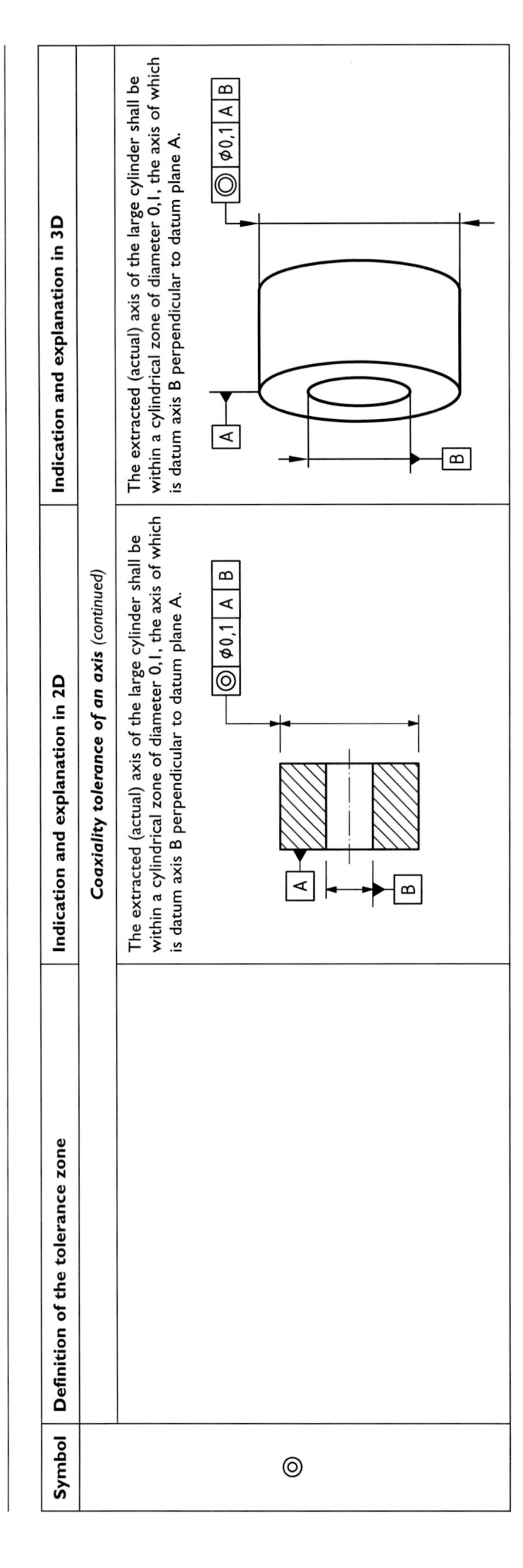

Symbol	Definition of the tolerance zone	Indication and explanation in 2D	Indication and explanation in 3D
		Coaxiality tolerance of an axis *(continued)*	
◎		The extracted (actual) axis of the large cylinder shall be within a cylindrical zone of diameter 0,1, the axis of which is datum axis B perpendicular to datum plane A. ⌖ ⌀0,1 A B	The extracted (actual) axis of the large cylinder shall be within a cylindrical zone of diameter 0,1, the axis of which is datum axis B perpendicular to datum plane A. ◎ ⌀0,1 A B

Geometric tolerancing datums and datum systems

Symbol	Definition of the tolerance zone	Indication and explanation in 2D	Indication and explanation in 3D
	The tolerance zone is limited by two parallel planes a distance t apart, symmetrically disposed about the median plane, with respect to the datum. a Datum.	**Symmetry tolerance of a median plane** The extracted (actual) median surface shall be contained between two parallel planes 0,08 apart which are symmetrically disposed about the datum median plane A. The extracted (actual) median surface shall be contained between two parallel planes 0,08 apart and symmetrically disposed about the common datum plane A-B.	The extracted (actual) median surface shall be contained between two parallel planes 0,08 apart which are symmetrically disposed about the datum median plane A. The extracted (actual) median surface shall be contained between two parallel planes 0,08 apart and symmetrically disposed about the common datum plane A-B.

103

Symbol	Definition of the tolerance zone	Indication and explanation in 2D	Indication and explanation in 3D
	The tolerance zone is limited within any cross-section perpendicular to the datum axis by two concentric circles with a difference in radii of t, the centres of which coincide with the datum. a Datum. b Cross-section plane	*Circular run-out tolerance — radial* The extracted (actual) line in any cross-section plane perpendicular to datum axis A shall be contained between two coplanar concentric circles with a difference in radii of 0,1.	The extracted (actual) line in any cross-section plane perpendicular to datum axis A shall be contained between two coplanar concentric circles with a difference in radii of 0,1.

Geometric tolerancing datums and datum systems

Symbol	Definition of the tolerance zone	Indication and explanation in 2D	Indication and explanation in 3D
↗		*Circular run-out tolerance — radial* (continued)	
		The extracted (actual) line in any cross-section plane parallel to datum plane B, shall be contained between two coplanar concentric circles concentric to datum axis A with a difference in radii of 0,1.	The extracted (actual) line in any cross-section plane parallel to datum plane B, shall be contained between two coplanar concentric circles concentric to datum axis A with a difference in radii of 0,1.

Symbol	Definition of the tolerance zone	Indication and explanation in 2D	Indication and explanation in 3D
↗		*Circular run-out tolerance — radial (continued)*	
		The extracted (actual) line in any cross-section plane perpendicular to common datum straight line A-B shall be contained between two coplanar concentric circles with a difference in radii of 0,1.	The extracted (actual) line in any cross-section plane perpendicular to common datum straight line A-B shall be contained between two coplanar concentric circles with a difference in radii of 0,1.

Geometric tolerancing datums and datum systems

Symbol	Definition of the tolerance zone	Indication and explanation in 2D	Indication and explanation in 3D
		Circular run-out tolerance — radial (continued)	
	Run-out usually applies to complete features, but could be applied to a restricted part of a feature (see 2D and 3D explanations).	The extracted (actual) line in any cross-section plane perpendicular to datum axis A shall be contained between two coplanar concentric circles with a difference in radii of 0,2.	The extracted (actual) line in any cross-section plane perpendicular to datum axis A shall be contained between two coplanar concentric circles with a difference in radii of 0,2.

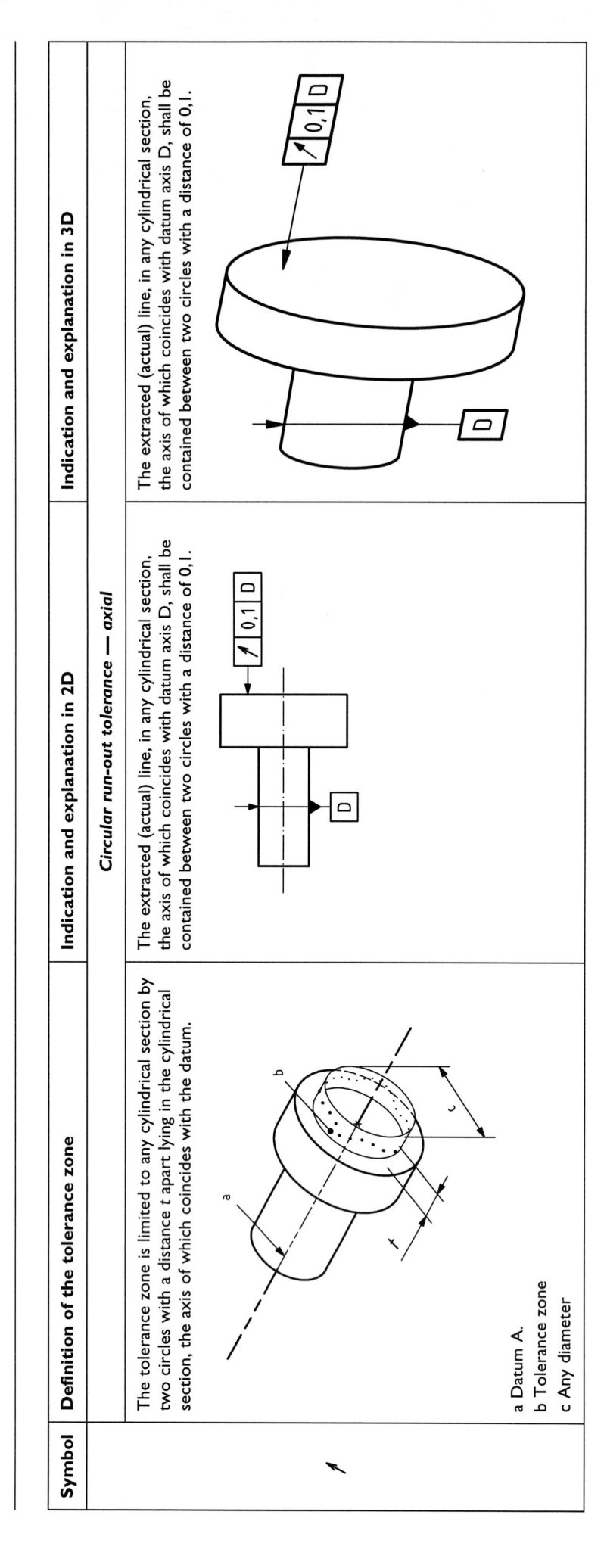

Symbol	Definition of the tolerance zone	Indication and explanation in 2D	Indication and explanation in 3D
	The tolerance zone is limited to any cylindrical section by two circles with a distance t apart lying in the cylindrical section, the axis of which coincides with the datum.	*Circular run-out tolerance — axial* The extracted (actual) line, in any cylindrical section, the axis of which coincides with datum axis D, shall be contained between two circles with a distance of 0,1.	The extracted (actual) line, in any cylindrical section, the axis of which coincides with datum axis D, shall be contained between two circles with a distance of 0,1.

a Datum A.
b Tolerance zone
c Any diameter

Geometric tolerancing datums and datum systems

Symbol	Definition of the tolerance zone	Indication and explanation in 2D	Indication and explanation in 3D
		Circular run-out tolerance in any direction	
↗	The tolerance zone is limited within any conical section by two circles a distance t apart, the axes of which coincide with the datum. The width of the tolerance zone is normal to the specified geometry unless otherwise indicated. a Datum C. b Tolerance zone.	The extracted (actual) line in any conical section, the axis of which coincides with datum axis C, shall be contained between two circles within the conical section with a distance of 0,1.	The extracted (actual) line in any conical section, the axis of which coincides with datum axis C, shall be contained between two circles within the conical section with a distance of 0,1.

Symbol	Definition of the tolerance zone	Indication and explanation in 2D	Indication and explanation in 3D
↗		*Circular run-out tolerance in any direction (continued)*	
		When the generator line for the toleranced feature is not straight, the apex angle of the conical section will change depending on the actual position [see definition (left column, page 109) and below].	When the generator line for the toleranced feature is not straight, the apex angle of the conical section will change depending on the actual position [see definition (left column) and below].

Geometric tolerancing datums and datum systems

Symbol	Definition of the tolerance zone	Indication and explanation in 2D	Indication and explanation in 3D
↗	The tolerance zone is limited within any conical section of the specified angle by two circles a distance t apart, the axes of which coincide with the datum. a Datum C. b Tolerance zone	*Circular run-out tolerance in a specified direction* The extracted (actual) line in any conical section (angle α), the axis of which coincides with datum axis C, shall be contained between two circles at a distance 0,1 apart within the conical section.	The extracted (actual) line in any conical section (angle α), the axis of which coincides with datum axis C, shall be contained between two circles at a distance 0,1 apart within the conical section.

Symbol	Definition of the tolerance zone	Indication and explanation in 2D	Indication and explanation in 3D
		Total radial run-out tolerance	
⟰	The tolerance zone is limited by two coaxial cylinders with a difference in radii of t, the axes of which coincide with the datum. a Datum A-B.	The extracted (actual) surface shall be contained between two coaxial cylinders with a difference in radii of 0,1 and the axes coincident with the common datum straight line A-B.	The extracted (actual) surface shall be contained between two coaxial cylinders with a difference in radii of 0,1 and the axes coincident with the common datum straight line A-B.

Geometric tolerancing datums and datum systems

113

Symbol	Definition of the tolerance zone	Indication and explanation in 2D	Indication and explanation in 3D
↗	The tolerance zone is limited by two parallel planes a distance *t* apart and perpendicular to the datum. a Datum D. b Extracted surface.	***Total axial run-out tolerance*** The extracted (actual) surface shall be contained between two parallel planes 0,1 apart which are perpendicular to datum axis D.	The extracted (actual) surface shall be contained between two parallel planes 0,1 apart which are perpendicular to the datum axis D.

2.11 Relevant standards

BS ISO 128-24, *Technical drawings — General principles of presentation — Part 24: Lines on mechanical engineering drawings*

BS EN ISO 1101, *Geometrical Product Specifications (GPS) — Geometrical tolerancing — Tolerances of form, orientation, location and run-out*

BS EN ISO 2692, *Geometrical product specifications (GPS) — Geometrical tolerancing — Maximum material requirement (MMR), least material requirement (LMR) and reciprocity requirement (RPR)*

BS EN ISO 5458, *Geometrical Product Specifications (GPS) — Geometrical tolerancing — Positional tolerancing*

BS ISO 5459, *Technical drawings — Geometrical tolerancing — Datums and datum-systems for geometrical tolerances*

BS ISO 8015, *Technical drawings — Fundamental tolerancing principle*

BS ISO 10579, *Technical drawings — Dimensioning and tolerancing — Non-rigid parts*

BS ISO 13715, *Technical drawings — Edges of undefined shape — Vocabulary and indications*

BS EN ISO 14660-1, *Geometrical Product Specifications (GPS) — Geometrical features — Part 1: General terms and definitions*

BS EN ISO 14660-2, *Geometrical Product Specifications (GPS) — Geometrical features — Part 2: Extracted median line of a cylinder and a cone, extracted median surface, local size of an extracted feature*

BS ISO 16792, *Technical product documentation — Digital product definition data practices*

Chapter 3

Graphical symbols for the indication of surface texture

3.1 Introduction

Surface texture requirements shall be indicated on technical product documentation by the use of several variants of a graphical symbol, each having its own significant meaning.

NOTE: For the purposes of this chapter, the definitions given in ISO 10209-1 and BS EN ISO 4287 are used.

3.2 The basic graphical symbol

The basic graphical symbol consists of two straight lines of unequal length inclined at approximately 60° to the line representing the considered surface, as shown in Figure 160. This symbol should not be used without complementary information specifying collective indications, as illustrated in Figure 181.

Figure 160 – Basic graphical symbol to indicate surface texture

3.3 Expanded graphical symbols

When removal of material (for example, by machining) is required, a horizontal bar is added to the basic symbol, as illustrated in Figure 161. This symbol indicates that a particular surface is machined but does not specify any surface texture.

Figure 161 – Graphical symbol to indicate removal of material by machining

When removal of material is not permitted, a circle is added to the basic symbol, as illustrated in Figure 162. This symbol indicates no material removal from a particular surface, but does not specify any surface texture. This symbol can also be used to indicate that a surface should remain in the same state resulting from a previous manufacturing process.

Figure 162 – Graphical symbol to indicate no removal of material

When complementary requirements for surface texture characteristics are specified, a horizontal line is added to the symbol, as illustrated in Figure 163.

Figure 163 – Graphical symbol to indicate surface texture characteristics

When the same surface texture is required on all surfaces around a workpiece, a circle is added to the graphical symbol, as shown in Figure 164.

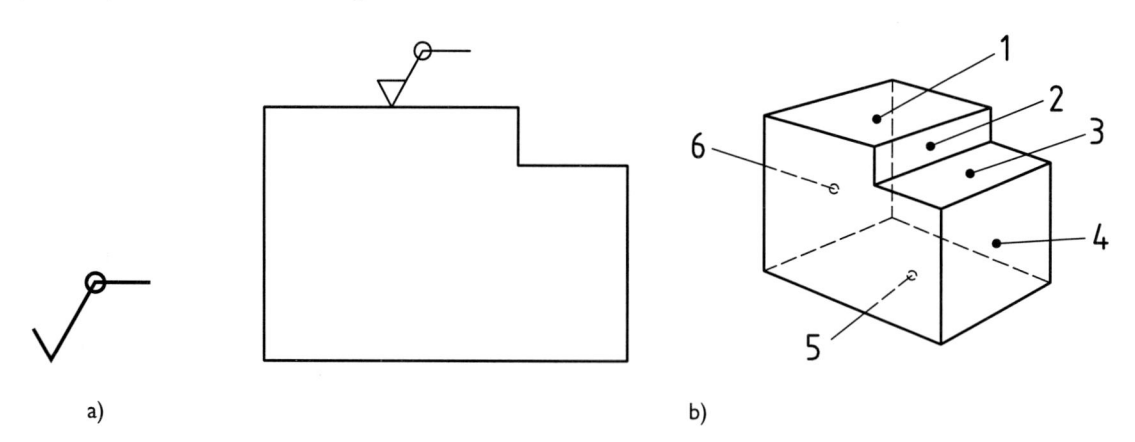

a) b)

Figure 164 – Graphical symbol to indicate the same surface texture is required on all surfaces around a workpiece

3.4 Mandatory positions for the indication of surface texture requirements

The indications of surface texture shall be placed relative to the graphical symbol as shown in Figure 165. Complementary surface texture requirements shall be in the form of:

- surface texture parameters;
- numerical values;
- sampling length/transmission band.

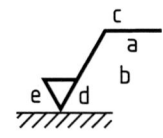

Figure 165 – Indications of surface texture relative to the graphical symbol

Key:

a – One single surface requirement. The surface texture parameter designation, the numerical value and the transmission band/sampling length should be indicated at position '**a**'. To avoid misinterpretation, a double space (double blank) should be inserted between the parameter designation and the limit value. Generally, the transmission band or sampling length should be indicated followed by an oblique stroke (/), followed by the surface texture parameter designation followed by its numerical value.

Example 1: 0,0025 – 0,8/Rz 6,8 (example with transmission band indicated).

Example 2: 0,8/Rz 6,8 (example with only sampling length indicated).

Example 3: 0,008 – 0.5/16/R 10.

NOTE: Generally, the transmission band is the wavelength range between two defined filters (see BS EN ISO 3274 and BS EN ISO 11562), and for the motif method is the wavelength range between two defined limits (see BS EN ISO 12085).

a, b – Two or more surface texture requirements. The first surface texture requirement should be indicated at position '**a**' and the second at position '**b**'. If a third or further requirement is to be indicated, the graphical symbol should be enlarged accordingly in the vertical direction, to make room for more lines. The positions '**a**' and '**b**' should be moved upwards, when the symbol is enlarged.

c – Manufacturing method. The manufacturing method, treatment, coatings or other requirements for the manufacturing process, etc., to produce the surface (e.g. turned, ground, plated) should be located at position '**c**'.

d – Surface lay and orientation. Indicate the symbol of the required surface lay and the orientation, if any, of the surface lay, e.g. '=', 'X' and 'M'.

e – Machining allowance. Indicate the required machining allowance, if any, as a numerical value in millimetres (see BS ISO 10135).

3.5 Surface texture parameters

Every surface of a workpiece has some form of texture, which varies according to the way it has been manufactured. Surface texture can be broken down into three main categories: surface roughness, waviness and form, which are defined as R, W and P profiles. The R profile series relates to roughness parameters, the W profile series to waviness parameters and the P profile series to form parameters.

Ra is the most used universally recognized international parameter of roughness. It is the arithmetic mean of the departures of the roughness profile from the mean, and applications of this parameter are illustrated in the various examples that follow.

Figures 166 to 169 illustrate the position of the chosen surface texture parameter value in conformation with 'a' and 'b' in Figure 165. When only one value is specified, it constitutes the upper limit of the surface roughness parameter. If it is necessary to specify upper and lower limits of the roughness parameter, both values should be given as illustrated in Figure 169, with the upper limit a1 above the lower limit a2.

Figure 166 – Surface texture parameter value added to basic graphical symbol

Figure 167 – Surface texture parameter value added to symbol for removal of material by machining

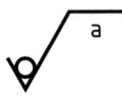

Figure 168 – Surface texture parameter value added to symbol for no material to be removed

Figure 169 – Upper and lower surface texture parameter values added to basic graphical symbol

3.6 Indication of special surface texture characteristics

In certain circumstances, for functional reasons, it may be necessary to specify additional special requirements concerning surface texture.

When specifying how the surface texture is to be produced, that method shall be indicated in words (see 'c' in Figure 165) on a line added to the longer arm of the symbols given in Figures 160 to 162, as shown in Figure 170.

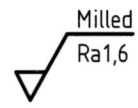

Figure 170 – Method of producing surface texture indicated in words on graphical symbol

Any indications relating to treatment or coatings should also be given on this line. Unless otherwise stated, the numerical value of the roughness applies to the surface texture after treatment or coating. If it is necessary to define surface texture both before and after treatment, this should be explained in a note or in accordance with Figure 171.

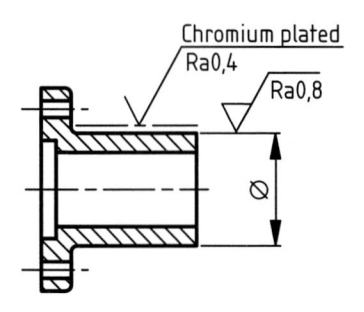

Figure 171 – Treatment or coatings to surface texture on graphical symbol

If it is necessary to indicate the sampling length, this should be selected from the appropriate series given in BS EN ISO 4288 and stated, in millimetres, adjacent to the graphical symbol, as shown in Figure 172. If it is necessary to specify the surface lay by working (e.g. tool marks) the symbol should be added to the surface texture symbol, as shown for example in Figure 173. The graphical symbols for the common surface patterns are specified in Table 5.

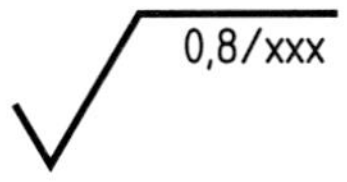

Figure 172 – Indication of sampling length on graphical symbol

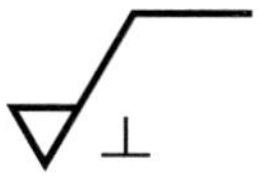

Figure 173 – Indication of surface lay by working on graphical symbol

Table 5 – Graphical symbols for common surface patterns

Graphical symbol [a]	Interpretation	Example
=	Parallel to the plane of projection of the view in which the symbol is used	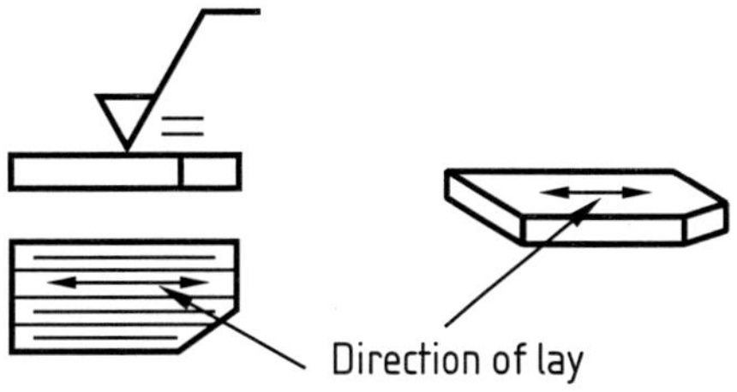 Direction of lay
⊥	Perpendicular to the plane of projection of the view in which the symbol is used	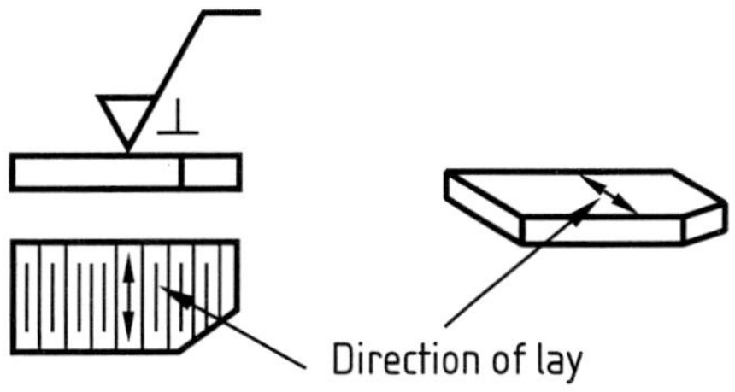 Direction of lay
X	Crossed in two oblique directions relative to the plane of projection of the view in which the symbol is used	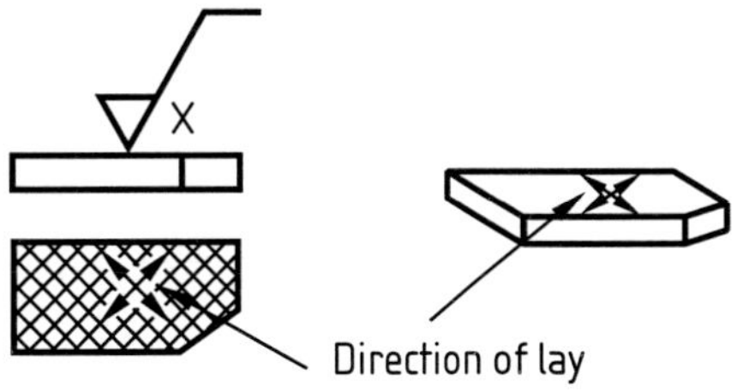 Direction of lay
M	Multi-directional	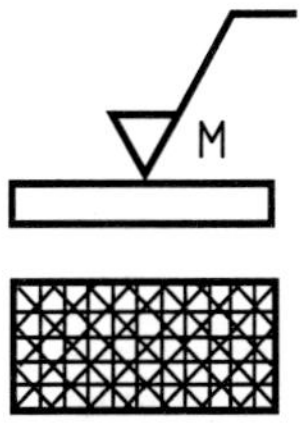

Graphical symbol[a]	Interpretation	Example
C	Approximately circular relative to the centre of the surface to which the symbol applies	
R	Approximately radial relative to the centre of the surface to which the symbol applies	
P	Lay is particulate, non-directional or protuberant	

[a] If it is necessary to specify a surface pattern that is not clearly defined by these symbols, this should be achieved by the addition of a suitable note to the drawing

When indicating a machining allowance, the relative information is positioned on the symbol as shown by 'e' in Figure 165. This allowance is generally indicated only where process stages are shown on the same drawing, i.e. on drawings of raw cast and forged workpieces, with the final workpiece depicted superimposed in the raw workpiece.

3.7 Indications on drawings

The general rule is that the graphical symbol together with the associated indications should be oriented so that they can be read from the bottom or the right-hand side of the drawing, as shown in Figure 174.

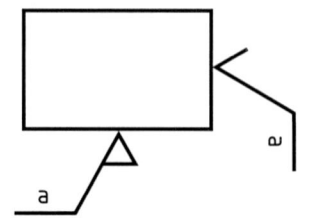

Figure 174 – Orientation of graphical symbols in relation to drawing views

However, if it is not practicable to adopt this general rule, the graphical symbol may be drawn in any position, but only if it does not carry any indications of special surface texture characteristics. Nevertheless, in such cases, the inscription defining the value of the arithmetical mean deviation 'Ra' (if present) should always be written in conformity with the general rule, as shown in Figure 174.

If necessary, the graphical symbol may be connected to the surface by a leader line terminating in an arrowhead.

As a general rule, the graphical symbol, or the leader line terminating in an arrowhead, shall point from outside the material of the piece, either to the line representing the surface or to an extension of it, as shown in Figure 175. However, if there is no risk of misinterpretation, the surface roughness requirement may be indicated in connection with the dimensions given, as shown in Figure 176.

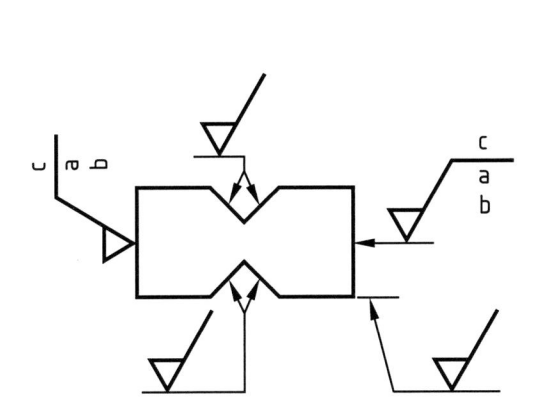

Figure 175 – Graphical symbol connected to a surface by a leader line

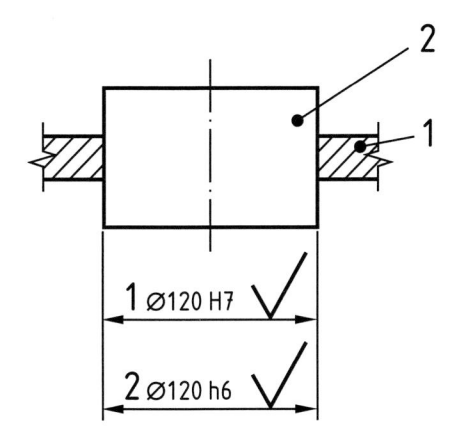

Figure 176 – Surface roughness requirement indicated in connection with dimensions

The graphical symbol should be used only once for a given surface and, if possible, on the same view as the dimensions defining the size or position of the surface. Cylindrical as well as prismatic surfaces shall only be specified once if indicated by a centre-line, as shown in Figure 177. However, each prismatic surface shall be indicated separately if a different surface texture is required or if particular requirements are applicable, see Figure 178.

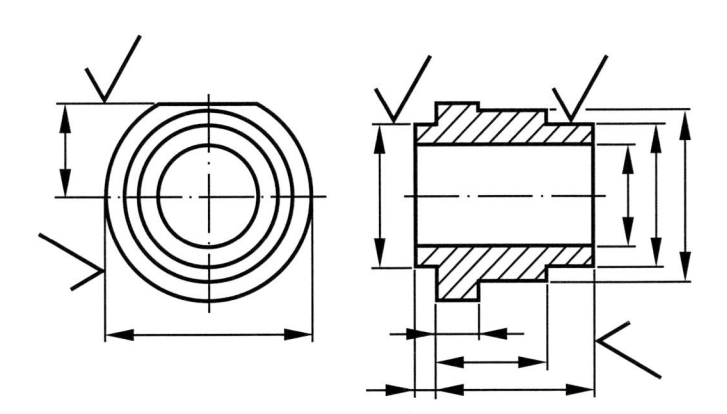

Figure 177 – Graphical symbol used only once for a given surface

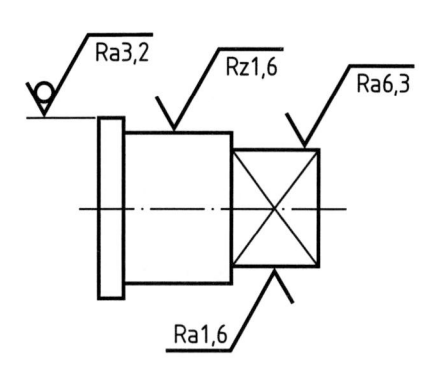

Figure 178 – Separate indication of each prismatic surface

If the same surface texture is required on the majority of the surfaces of a part, this surface texture requirement should be placed near the title block or in a space provided in the title block of a drawing.
The general graphical symbol corresponding to this surface texture should be followed by:

- a basic graphical symbol in parentheses without any other indication as shown in Figure 179; or
- the graphical symbol or symbols in parentheses of the special deviating surface texture requirements in order to indicate requirements that deviate from the general surface texture requirement, as shown in Figure 179.

Symbols for surface textures that are exceptions to the general symbol should be indicated on the corresponding surfaces.

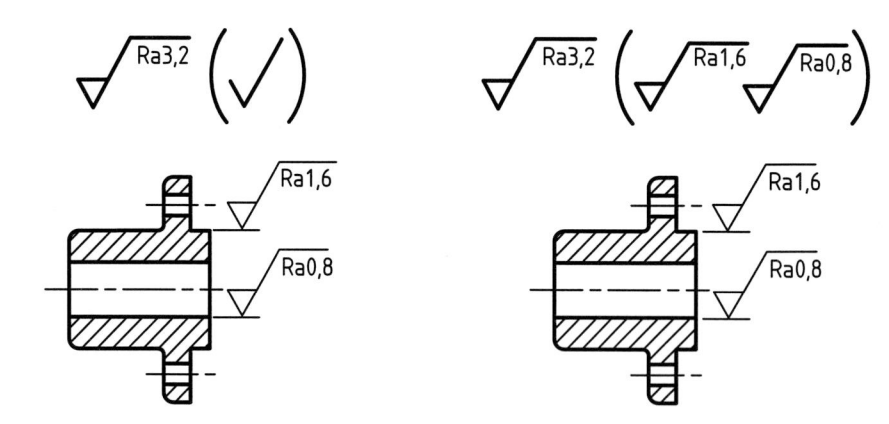

Figure 179 – Indication of the same surface texture on the majority of surfaces of a part

To avoid the necessity of repeating a complicated indication a number of times, or where space is limited, a simplified indication may be used on the surface, provided that its specification is explained near the part in question, near the title block or in the space devoted to general notes, as shown in Figure 180.

Graphical symbols for the indication of surface texture

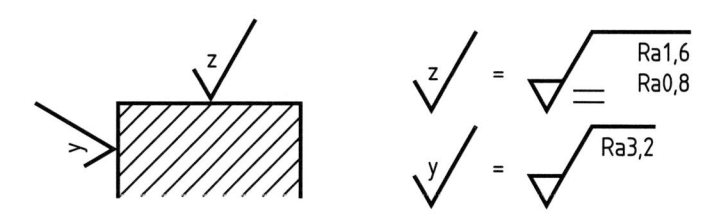

Figure 180 – Simplified indication of surface texture

If the same surface texture is required on a large number of surfaces of the part, the corresponding graphical symbol shown in Figures 160 to 162 may be used on the appropriate surface and its specification given on the drawing, near the title block or in the space devoted to general notes, as shown in Figure 181.

Figure 181 – Simplified indication of surface texture

3.8 Relevant standards

BS EN ISO 1302, *Geometrical Product Specifications (GPS) — Indication of surface texture in technical product documentation*

BS EN ISO 3274, *Geometric Product Specifications (GPS) — Surface texture: Profile method — Nominal characteristics of contact (stylus) instruments*

BS EN ISO 4287, *Geometrical product specification (GPS) — Surface texture: Profile method — Terms, definitions and surface texture parameters*

BS EN ISO 4288, *Geometric Product Specification (GPS) — Surface texture — Profile method: Rules and procedures for the assessment of surface texture*

BS EN ISO 8785, *Geometrical Product Specification (GPS) — Surface imperfections — Terms, definitions and parameters*

BS ISO 10135, *Geometrical product specifications (GPS) — Drawing indications for moulded parts in technical product documentation (TPD)*

BS ISO 10209-1, *Technical product documentation — Vocabulary — Terms relating to technical drawings: general and types of drawings*

BS EN ISO 11562, *Geometric product specifications (GPS) — Surface texture: Profile method. Metrological characteristics of phase correct filters*

BS EN ISO 12085, *Geometric Product Specification (GPS) — Surface texture: Profile method — Motif parameters*

Chapter 4

Welding, brazed and soldered joints – Symbolic representation

4.1 Introduction

The illustrations on pages 125–133 are taken from wall chart BS 499-2C:1999, *European Arc Welding Symbols*, which is based on BS EN 22553, *Welded, brazed and soldered joints — Symbolic representation on drawings*.

NOTE: All drawings where these symbols are used are to be referenced BS EN 22553 (ISO 2553:1992).

1. ELEMENTARY SYMBOLS

Type of weld	Illustration	Symbol
Butt weld between plates with raised edges which are melted down completely		
Square butt weld		
Single-V butt weld		
Single-bevel butt weld		
Single-U butt weld		

Type of weld	Illustration	Symbol
Single-J butt weld		⊩
Backing run		⌣
Double-V butt weld		X
Double-bevel butt weld		K
Double-U butt weld		⊃⊂
Fillet weld		◺
Plug weld (plug or slot weld — USA)		⊓
Surfacing		⌒

Welding, brazed and soldered joints – Symbolic representation

2. SUPPLEMENTARY SYMBOLS

Shape of weld surface or weld	Supplementary symbol
Flat (usually finished flush by grinding or machining)	———
Convex	⌒
Concave	⌣
Toes shall be blended smoothly — may require dressing	⏝
Permanent backing strip used	M
Removable backing strip used	MR

Examples of the use of supplementary symbols		
Designation	**Illustration**	**Symbol**
Flat (flush) single-V butt weld with permanent backing strip		
Flat (flush) single-V butt weld with flat (flush) backing run		
Convex double-V weld		
Concave fillet weld		
Fillet weld with toes smoothly blended		

3. REFERENCE LINES AND OTHER INFORMATION

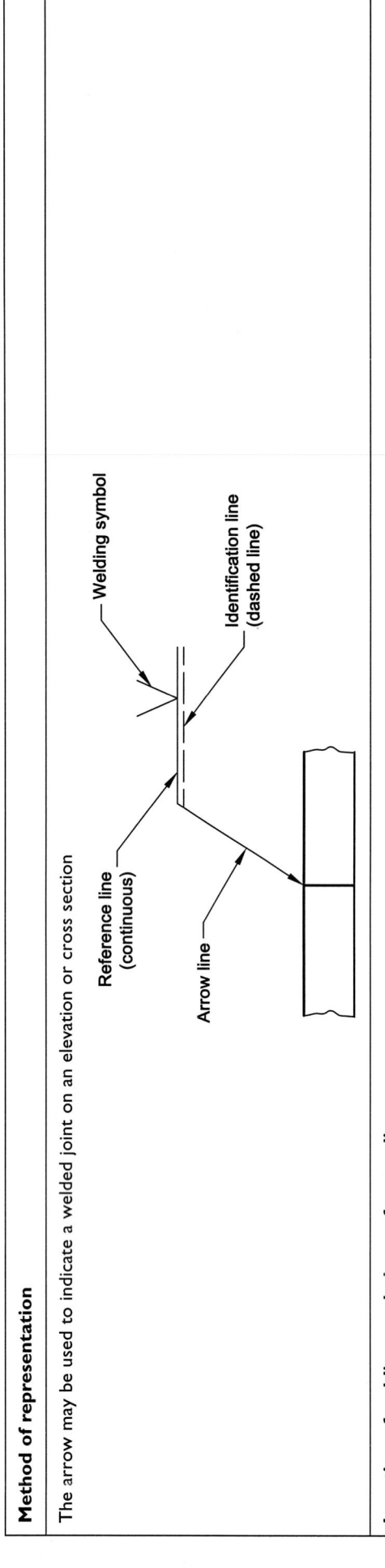

Method of representation

The arrow may be used to indicate a welded joint on an elevation or cross section

Location of welding symbol on reference line

It is recommended that the arrow line is placed on the side of the joint to be welded unless there is not enough space

It is recommended that the welding symbol is placed on the reference line but this is not mandatory

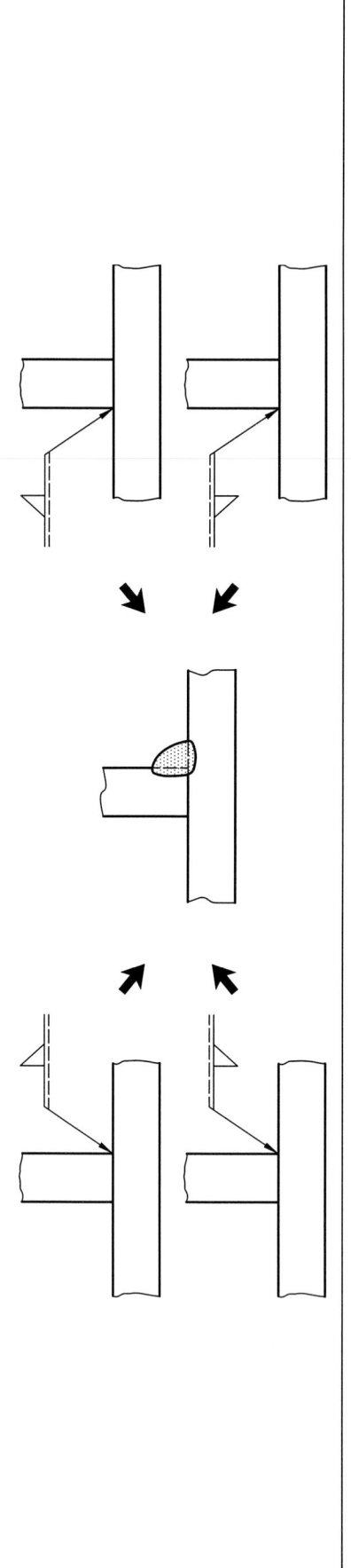

Special rules for butt welds

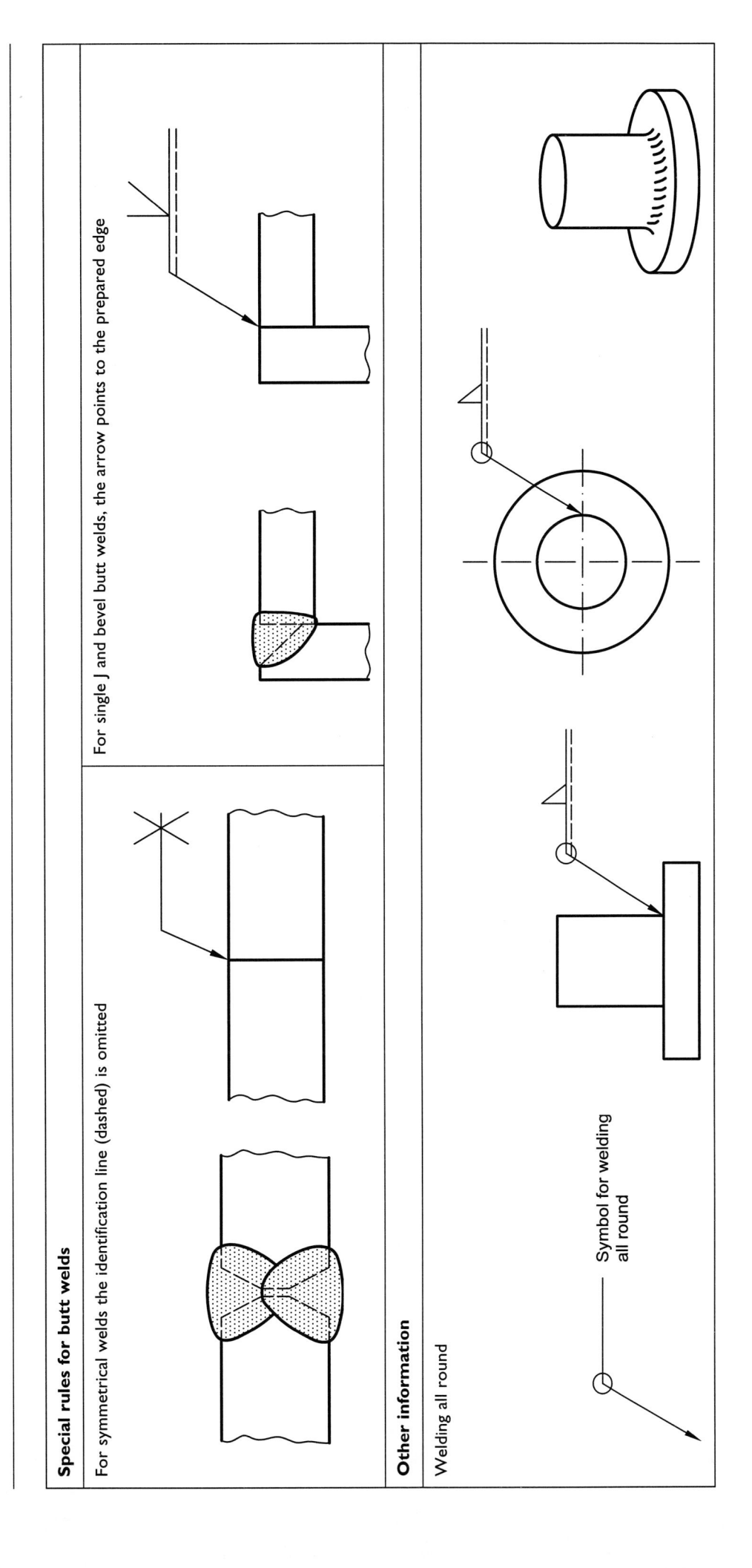

For symmetrical welds the identification line (dashed) is omitted

For single J and bevel butt welds, the arrow points to the prepared edge

Other information

Welding all round

Symbol for welding all round

Special rules for butt welds *(continued)*

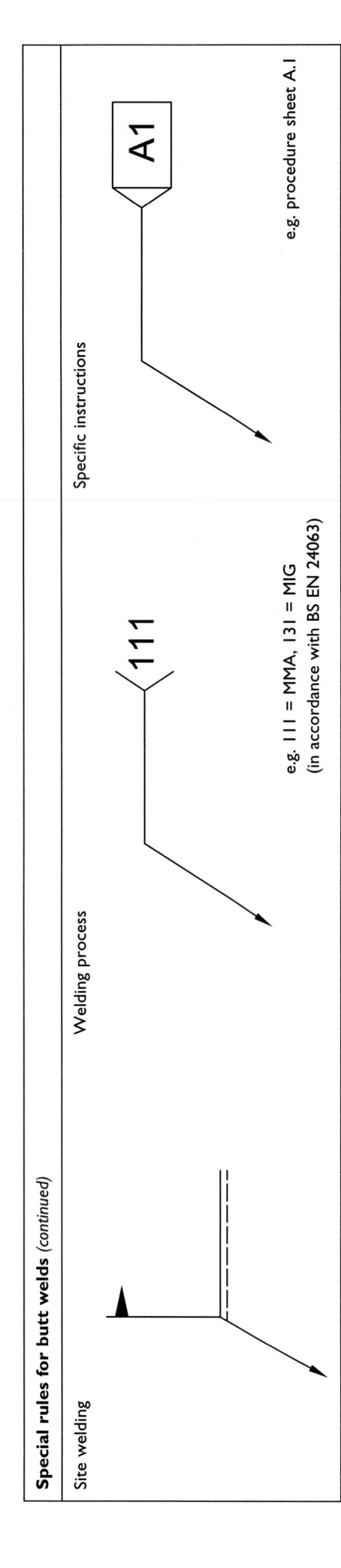

Site welding	Welding process	Specific instructions

e.g. 111 = MMA, 131 = MIG
(in accordance with BS EN 24063)

e.g. procedure sheet A.1

Welding, brazed and soldered joints – Symbolic representation

4. WELD DIMENSIONS

Butt welds

's' = minimum specified throat (penetration) thickness. If no dimension is shown, the weld is full penetration

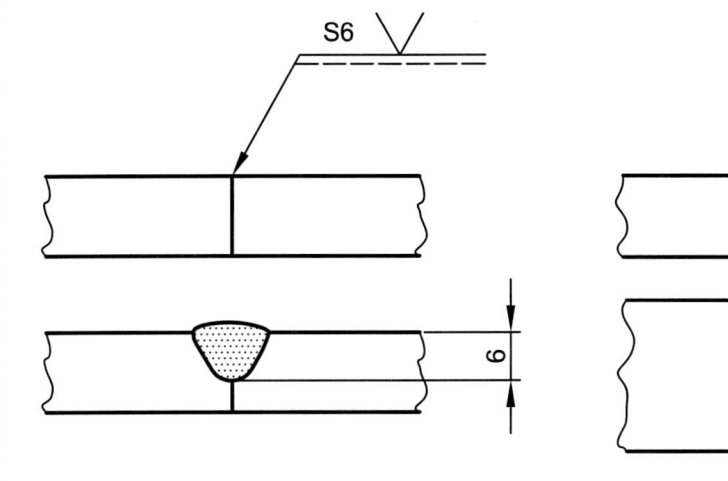

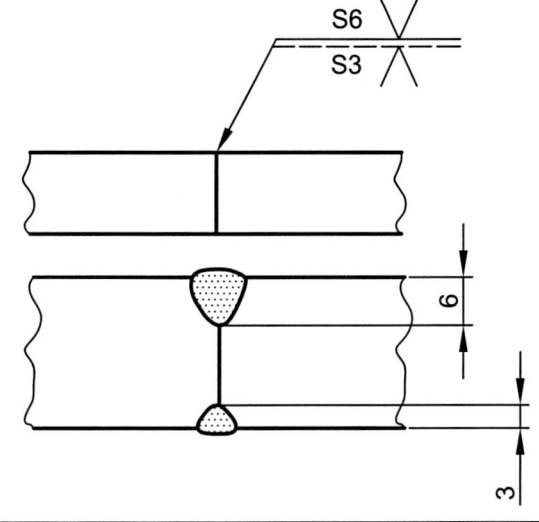

Fillet welds

'a' = throat thickness

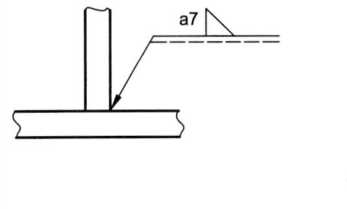

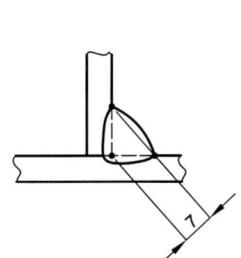

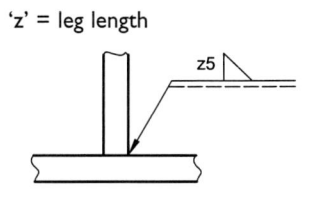

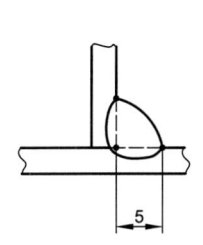

'z' = leg length

Deep penetration fillet welds

The throat thickness is designated by 's' and the dimensions are given for example 's8a6'

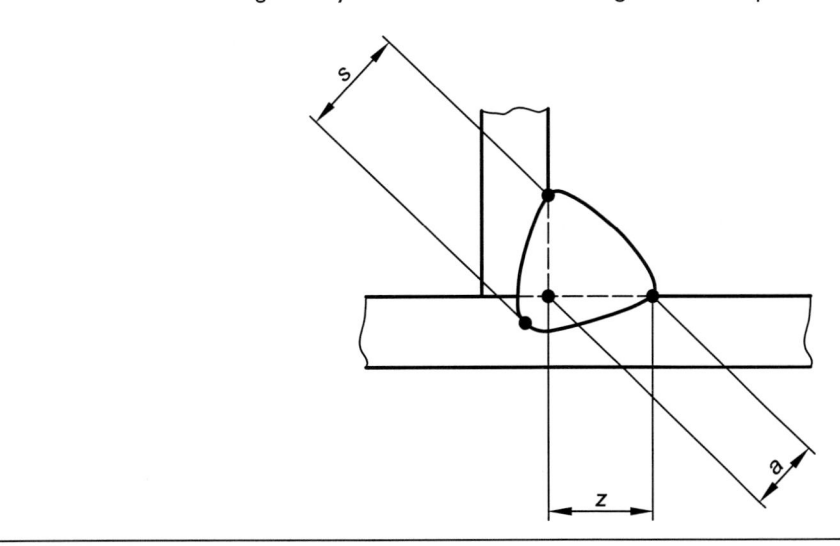

Weld length

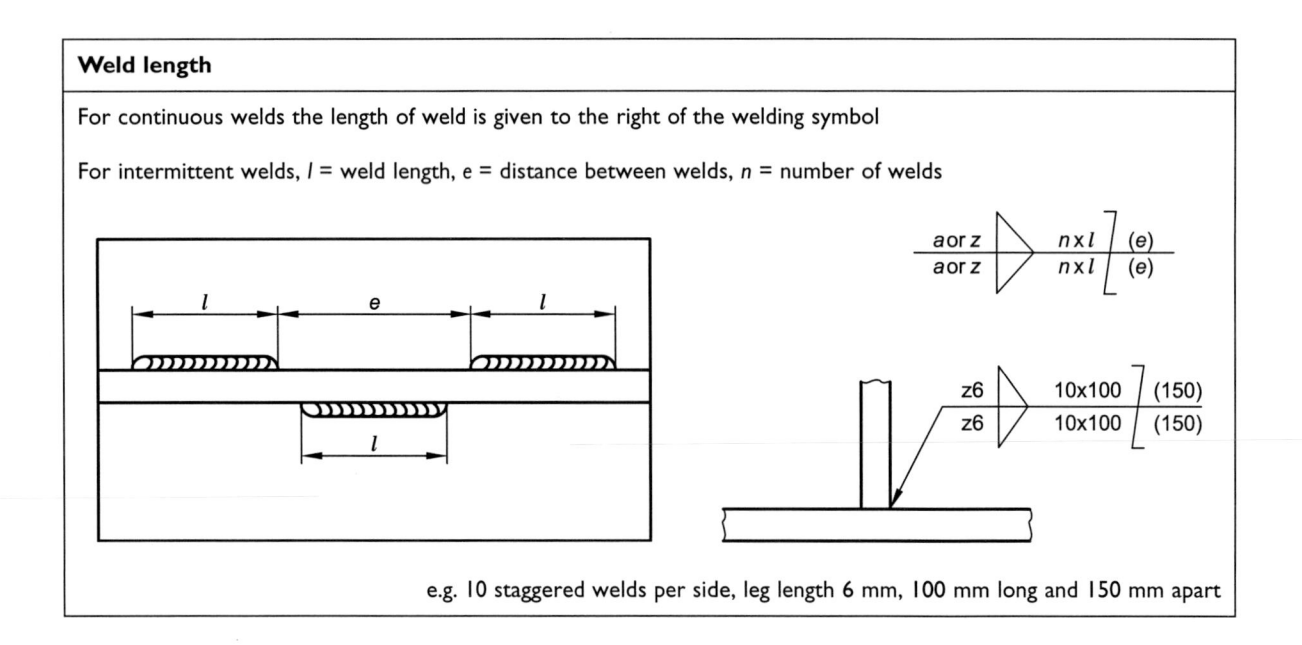

For continuous welds the length of weld is given to the right of the welding symbol

For intermittent welds, *l* = weld length, *e* = distance between welds, *n* = number of welds

e.g. 10 staggered welds per side, leg length 6 mm, 100 mm long and 150 mm apart

5. EXAMPLES SHOWING THE USE OF SYMBOLS

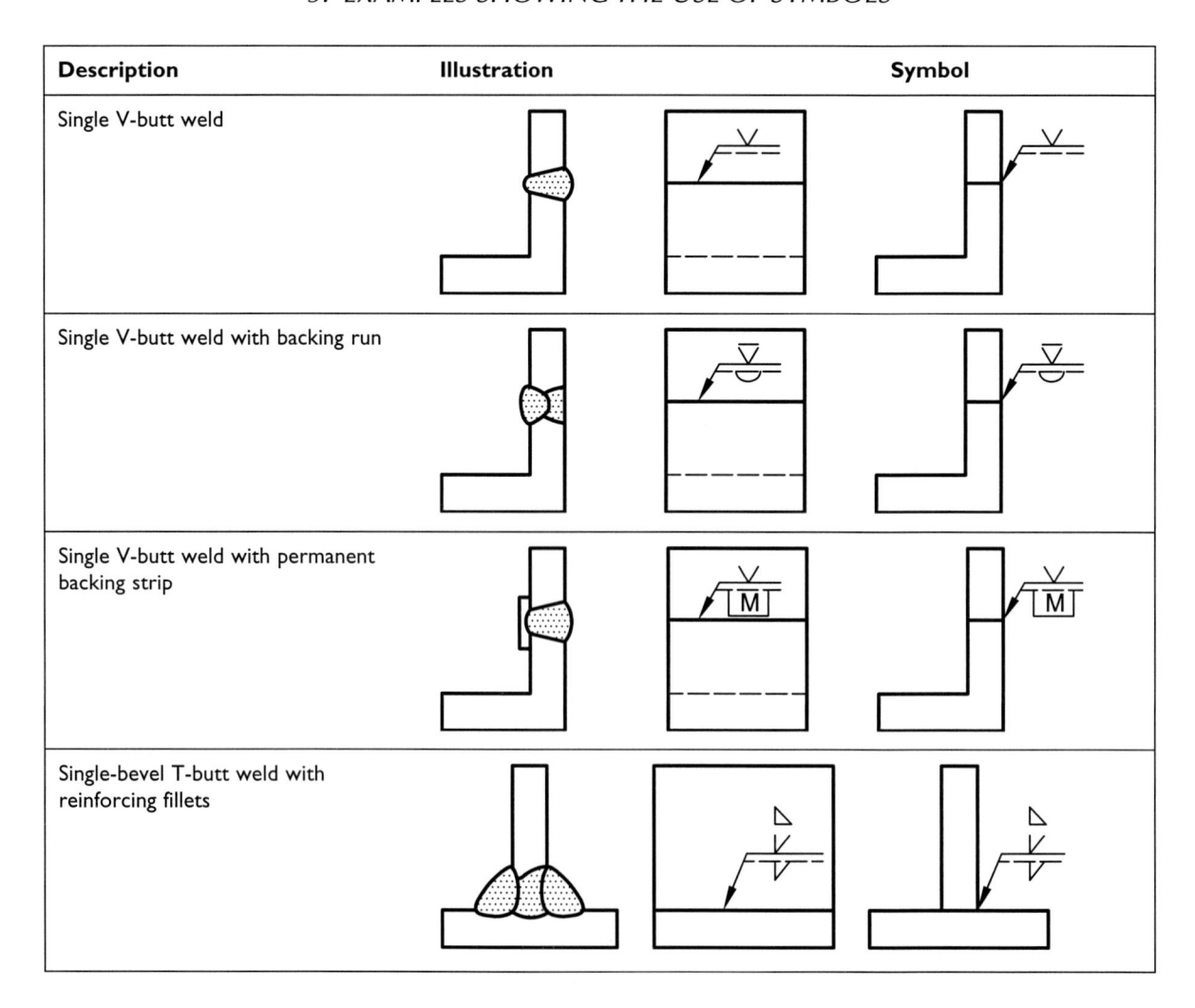

Description	Illustration	Symbol
Single V-butt weld		
Single V-butt weld with backing run		
Single V-butt weld with permanent backing strip		
Single-bevel T-butt weld with reinforcing fillets		

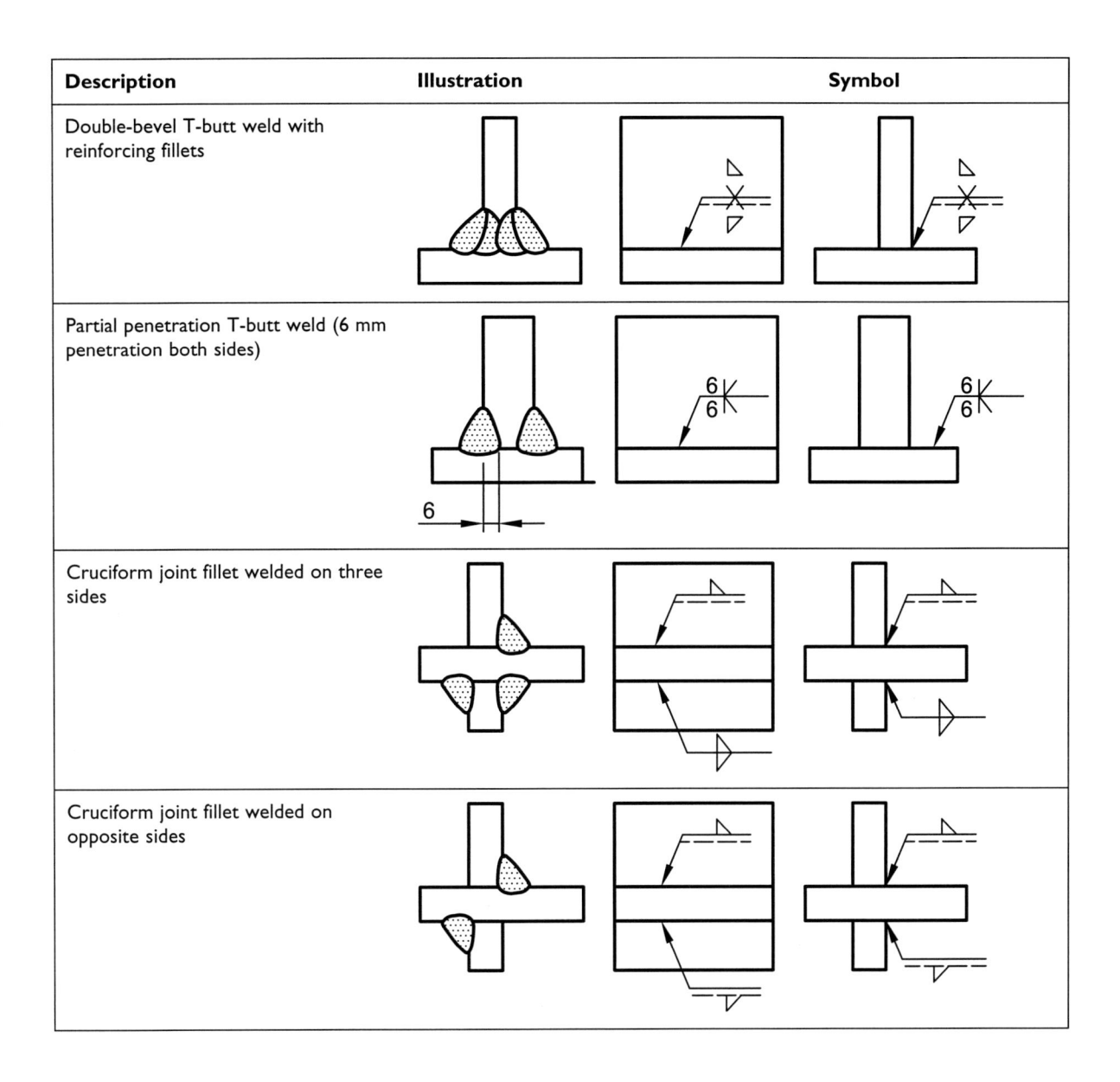

Welding, brazed and soldered joints – Symbolic representation

Description	Illustration	Symbol
Double-bevel T-butt weld with reinforcing fillets		
Partial penetration T-butt weld (6 mm penetration both sides)		
Cruciform joint fillet welded on three sides		
Cruciform joint fillet welded on opposite sides		

4.2 Relevant standards

BS 499-2c:1999, *European Arc Welding Symbols*
BS EN 22553:1995, *Welded, brazed and soldered joints — Symbolic representation on drawings*

Chapter 5

Limits and fits

5.1 Introduction

The interchangeability of mass produced parts requires a system of limits and fits to ensure a correct functional fit between two mating parts such as a shaft in a hole. This is achieved by specifying limits of size (tolerances) for each part (or a feature on a part). The maximum and minimum deviations in these sizes characterize the type of fit, i.e. clearance, transition and interference.

The term 'hole' or 'shaft' is used to designate features of size of a cylinder (tolerancing of a hole or shaft) or the size between two parallel surfaces (thickness of a key or width of a slot).

5.2 Selected ISO fits – Hole basis

The content of this section, including the table on page 137, is taken from BS 4500A, *Selected ISO Fits – hole basis*.

The ISO system provides a great many hole and shaft tolerances so as to cater for a very wide range of conditions. However, experience shows that the majority of fit conditions required for normal engineering products can be provided by a quite limited selection of tolerances.

The following selected hole and shaft tolerances have been found to be commonly applied:
Selected hole tolerances: H7; H8; H9; H11
Selected shaft tolerances: c11; d10; e9; f7; g6; h6; k6; n6; p6; s6

The table in this data sheet shows a range of fits derived from these selected hole and shaft tolerances. As will be seen, it covers fits from loose clearance to heavy interference and it may therefore be found to be suitable for most normal requirements. Many users may in fact find that their needs are met by a further selection within this selected range.

It should be noted, however, that this table is offered only as an example of how a restricted selection of fits can be made. It is clearly impossible to recommend selection of fits which are appropriate to all sections of industry, but it must be emphasized that a user who decides upon a selected range will always enjoy the economic advantages this conveys. Once he has installed the necessary tooling and gauging facilities, he can combine his selected hole and shaft tolerances in different ways without any additional investment in tools and equipment.

For example, if it is assumed that the range of fits shown in the table has been adopted but that, for a particular application the fit H8–f7 is appropriate but provides rather too much variation, the hole tolerance H7 could equally well be associated with the shaft f7 and may provide exactly what is required without necessitating any additional tooling.

For most general applications it is usual to recommend hole basis fits as, except in the realm of very large sizes where the effects of temperature play a large part, it is usually considered easier to manufacture and measure the male member of a fit and it is thus desirable to be able to allocate the larger part of the tolerance available to the hole and adjust the shaft to suit.

In some circumstances, however, it may in fact be preferable to employ a shaft-basis. For example, in the case of driving shafts where a single shaft may gave to accommodate a variety of accessories such as couplings, bearings, collars, etc., it is preferable to maintain a constant diameter for the permanent member, which is the shaft, and vary the bore of the accessories. For use in applications of this kind, a selection of shaft basis fits is provided in Data Sheet 4500B.

Limits and fits

SELECTED ISO FITS—HOLE BASIS
BRITISH STANDARD

Extracted from BS 4500 : 1969

Data Sheet 4500A
Issue 1. February 1970
Confirmed August 1985

5.3 Selected ISO fits – Shaft basis

The content of this section, including the table on page 139, is taken from BS 4500B, *Selected ISO Fits – shaft basis*.

The ISO system provides a great many hole and shaft tolerances so as to cater for a very wide range of conditions. However, experience shows that the majority of fit conditions required for normal engineering products can be provided by a quite limited selection of tolerances.

The following selected hole and shaft tolerances have been found to be commonly applied:
Selected hole tolerances: H7; H8; H9; H11
Selected shaft tolerances: c11; d10; e9; f7; g6; h6; k6; n6; p6; s6

For most general applications it is usual to recommend hole basis fits, i.e. fits in which the design size for the hole is the basic size and variations in the grade of fit for any particular hole are obtained by varying the clearance and the tolerance on the shaft. Data Sheet 4500A gives a range of hole basis fits derived from the selected hole and shaft tolerances above.

In some circumstances, however, it may in fact be preferable to employ a shaft basis. For example, in the case of driving shafts where a single shaft may have to accommodate a variety of accessories such as couplings, bearings, collars, etc., it is preferable to maintain a constant diameter for the permanent member, which is the shaft, and vary the bore of the accessories. Shaft basis fits also provide a useful economy where bar stock material is available to standard shaft tolerances of the ISO system.

For the benefit of those wishing to use shaft basis fits, this data sheet shows the shaft basis equivalents of the hole basis fits in Data Sheet 4500A. They are all direct conversions except that the fit H9–d10, instead of being converted to D9–h10, is adjusted to D10–h9 to avoid introducing the additional shaft tolerance h10.

As will be seen, the table covers fits from loose clearance to heavy interference and may therefore be found suitable for most normal requirements. Many users may in fact find that their needs are met by a further selection within this selected range.

It should be noted, however, that this Table is offered only as an example of how a restricted selection of fits can be made. It is clearly impossible to recommend selections of fits which are appropriate to all sections of industry, but it must be emphasized that a user who decides upon a selected range will always enjoy the economic advantages this conveys. Once he has installed the necessary tooling and gauging facilities, he can combine his selected hole and shaft tolerances in different ways without any additional investment in tools and equipment.

For example, if it is assumed that the range of fits shown in the table has been adopted but that, for a particular application the fit F8–h7 is appropriate but provides rather too much variation, the shaft tolerance h6 could equally well be associated with the hole F8 and may provide exactly what is required without necessitating any additional tooling.

Limits and fits

BRITISH STANDARD
SELECTED ISO FITS—SHAFT BASIS

Extracted from BS 4500 : 1969

Data Sheet 4500B
Issue 1. February 1970

		Clearance fits											Transition fits						Interference fits				Nominal sizes	
Nominal sizes		Tolerance		Tolerance		Tolerance		Tolerance		Tolerance		Tolerance		Tolerance		Tolerance		Tolerance		Tolerance		Tolerance		
Over	To	h11	C11	h9	D10	h9	E9	h7	F8	h6	G7	h6	H7	h6	K7	h6	N7	h6	P7	h6	S7	Over	To
mm	mm	0.001 mm	0.001 mm	0.001 mm	0.001 mm	0.001 mm	0.001 mm	0.001 mm	0.001 mm	0.001 mm	0.001 mm	0.001 mm	0.001 mm	0.001 mm	0.001 mm	0.001 mm	0.001 mm	0.001 mm	0.001 mm	0.001 mm	0.001 mm	mm	mm
—	3	0 / −60	+120 / +60	0 / −25	+60 / +20	0 / −25	+39 / +14	0 / −10	+20 / +6	0 / −6	+12 / +2	0 / −6	+10 / 0	0 / −6	0 / −10	0 / −6	−4 / −14	0 / −6	−6 / −16	0 / −6	−14 / −24	—	3
3	6	0 / −75	+145 / +70	0 / −30	+78 / +30	0 / −30	+50 / +20	0 / −12	+28 / +10	0 / −8	+16 / +4	0 / −8	+12 / 0	0 / −8	+3 / −9	0 / −8	−4 / −16	0 / −8	−8 / −20	0 / −8	−15 / −27	3	6
6	10	0 / −90	+170 / +80	0 / −36	+98 / +40	0 / −36	+61 / +25	0 / −15	+35 / +13	0 / −9	+20 / +5	0 / −9	+15 / 0	0 / −9	+5 / −10	0 / −9	−4 / −19	0 / −9	−9 / −24	0 / −9	−17 / −32	6	10
10	18	0 / −110	+205 / +95	0 / −43	+120 / +50	0 / −43	+75 / +32	0 / −18	+43 / +16	0 / −11	+24 / +6	0 / −11	+18 / 0	0 / −11	+6 / −12	0 / −11	−5 / −23	0 / −11	−11 / −29	0 / −11	−21 / −39	10	18
18	30	0 / −130	+240 / +110	0 / −52	+149 / +65	0 / −52	+92 / +40	0 / −21	+53 / +20	0 / −13	+28 / +7	0 / −13	+21 / 0	0 / −13	+6 / −15	0 / −13	−7 / −28	0 / −13	−14 / −35	0 / −13	−27 / −48	18	30
30	40	0 / −160	+280 / +120	0 / −62	+180 / +80	0 / −62	+112 / +50	0 / −25	+64 / +25	0 / −16	+34 / +9	0 / −16	+25 / 0	0 / −16	+7 / −18	0 / −16	−8 / −33	0 / −16	−17 / −42	0 / −16	−34 / −59	30	40
40	50	0 / −160	+290 / +130																			40	50
50	65	0 / −190	+330 / +140	0 / −74	+220 / +100	0 / −74	+134 / +60	0 / −30	+76 / +30	0 / −19	+40 / +10	0 / −19	+30 / 0	0 / −19	+9 / −21	0 / −19	−9 / −39	0 / −19	−21 / −51	0 / −19	−42 / −72	50	65
65	80	0 / −190	+340 / +150																		−48 / −78	65	80
80	100	0 / −220	+390 / +170	0 / −87	+260 / +120	0 / −87	+159 / +72	0 / −35	+90 / +36	0 / −22	+47 / +12	0 / −22	+35 / 0	0 / −22	+10 / −25	0 / −22	−10 / −45	0 / −22	−24 / −59	0 / −22	−58 / −93	80	100
100	120	0 / −220	+400 / +180																		−66 / −101	100	120
120	140	0 / −250	+450 / +200	0 / −100	+305 / +145	0 / −100	+185 / +85	0 / −40	+106 / +43	0 / −25	+54 / +14	0 / −25	+40 / 0	0 / −25	+12 / −28	0 / −25	−12 / −52	0 / −25	−28 / −68	0 / −25	−77 / −117	120	140
140	160	0 / −250	+460 / +210																		−85 / −125	140	160
160	180	0 / −250	+480 / +230																		−93 / −133	160	180
180	200	0 / −290	+530 / +240	0 / −115	+355 / +170	0 / −115	+215 / +100	0 / −46	+122 / +50	0 / −29	+61 / +15	0 / −29	+46 / 0	0 / −29	+13 / −33	0 / −29	−14 / −60	0 / −29	−33 / −79	0 / −29	−105 / −151	180	200
200	225	0 / −290	+550 / +260																		−113 / −159	200	225
225	250	0 / −290	+570 / +280																		−123 / −169	225	250
250	280	0 / −320	+620 / +300	0 / −130	+400 / +190	0 / −130	+240 / +110	0 / −52	+137 / +56	0 / −32	+62 / +17	0 / −32	+52 / 0	0 / −32	+16 / −36	0 / −32	−14 / −66	0 / −32	−36 / −88	0 / −32	−138 / −190	250	280
280	315	0 / −320	+650 / +330																		−150 / −202	280	315
315	355	0 / −360	+720 / +360	0 / −140	+440 / +210	0 / −140	+265 / +125	0 / −57	+151 / +62	0 / −36	+75 / +18	0 / −36	+57 / 0	0 / −36	+17 / −40	0 / −36	−16 / −73	0 / −36	−41 / −98	0 / −36	−169 / −226	315	355
355	400	0 / −360	+760 / +400																		−187 / −244	355	400
400	450	0 / −400	+840 / +440	0 / −155	+480 / +230	0 / −155	+290 / +135	0 / −63	+165 / +68	0 / −40	+83 / +20	0 / −40	+63 / 0	0 / −40	+18 / −45	0 / −40	−17 / −80	0 / −40	−45 / −108	0 / −40	−209 / −272	400	450
450	500	0 / −400	+880 / +480																		−229 / −292	450	500

5.4 Methods of specifying required fits

Fits taken from BS 4500-1 and BS 4500-2 shall be designated on drawings by the methods shown in Figures 182 and 183.

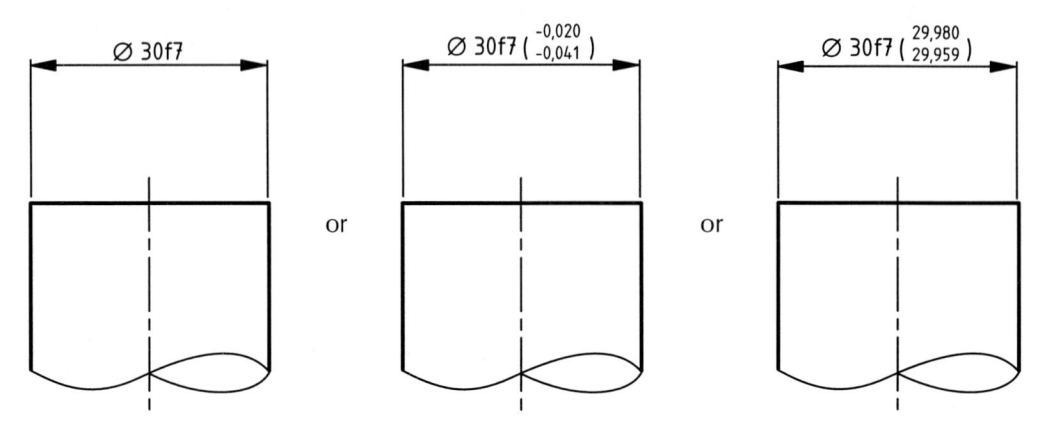

Figure 182 – Fit designation – shafts

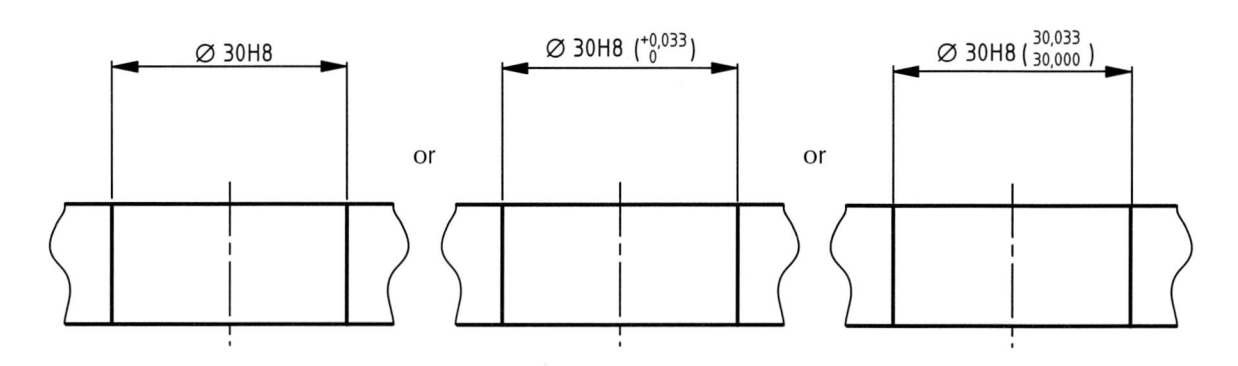

Figure 183 – Fit designation – holes

5.5 Relevant standards

BS 4500A Data Sheet: *Selected ISO fits – hole basis*
BS 4500B Data Sheet: *Selected ISO fits – shaft basis*
BS EN 20286-1, *ISO system of limits and fits — Part 1: Bases of tolerances, deviations and fits*
BS EN 20286-2, *ISO system of limits and fits — Part 2: Tables of standard tolerance grades and limit deviations for holes and shafts*

Chapter 6

Metric screw threads

6.1 Introduction

The table on pages 143–167 has been extracted from BS 3643-2, *ISO metric screw threads — Part 2: Specification for selected limits of size*. It specifies the fundamental deviations, tolerances and limits of size for the tolerance classes 4H, 5H, 6H and 7H for internal threads and 4h, 5h, 6h and 7h for external threads for:

- the coarse pitch series, ranging from 1 mm to 68 mm diameter;
- the fine pitch series, ranging from 1 mm to 33 mm diameter;
- the constant pitch series, ranging from 8 mm to 125 mm diameter.

For constant pitch series between 125 mm diameter and 300 mm diameter, see BS 3643-2.

6.2 Thread designation

The complete designation for a screw thread shall comprise a thread system and tolerance class. The letter M signifies the metric thread system.

Examples

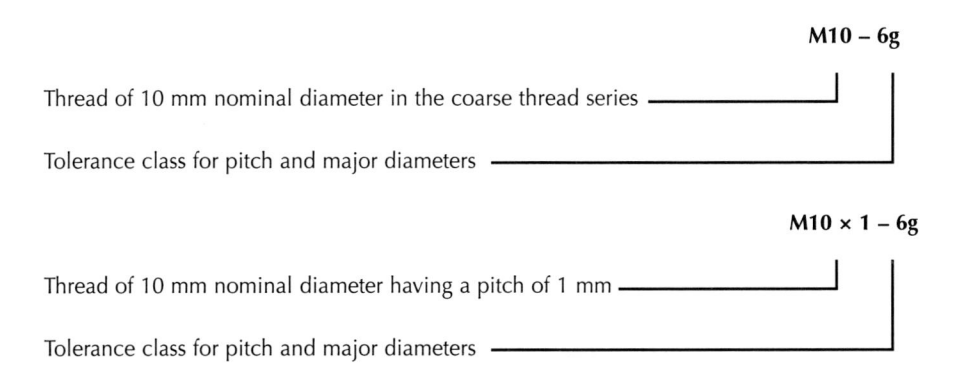

External thread

M10 – 6g

Thread of 10 mm nominal diameter in the coarse thread series

Tolerance class for pitch and major diameters

M10 × 1 – 6g

Thread of 10 mm nominal diameter having a pitch of 1 mm

Tolerance class for pitch and major diameters

Internal thread

          M10 – 6H

Thread of 10 mm nominal diameter in the coarse thread series ⏋

Tolerance class for pitch and major diameters ⏌

          M10 × 1 – 6H

Thread of 10 mm nominal diameter having a pitch of 1 mm ⏋

Tolerance class for pitch and major diameters ⏌

NOTE: In the absence of a specified pitch in the thread designation, the default is the coarse thread series.

The following table gives the limits and tolerances for finished, uncoated ISO metric screw threads for normal length of engagement.

Table 1 ISO metric screw threads – Limits and tolerances for finished uncoated threads for normal length of engagement

All dimensions are in millimetres.

1	2	3	4	5	6	7	8	9	10	11	12	13	14	15	16	17	18	19	20	21	22
Nominal diameter	Pitch			External threads									Internal threads								
	Coarse	Fine	Constant	Tolerance class	Fund dev.	Major diameter			Pitch diameter			Minor diameter	Tolerance class	Fund dev.	Major diameter	Pitch diameter			Minor diameter		
						max.	tol.	min.	max.	tol.	min.	min.			min.	max.	tol.	min.	max.	tol.	min.
1		0.2		4h	0	1.000	0.036	0.964	0.870	0.030	0.840	0.717	4H	0	1.000	0.910	0.040	0.870	0.821	0.038	0.783
				6g	0.017	0.983	0.056	0.927	0.853	0.048	0.805	0.682									
	0.25			4h	0	1.000	0.042	0.958	0.838	0.034	0.804	0.649	4H	0	1.000	0.883	0.045	0.838	0.774	0.045	0.729
				6g	0.018	0.982	0.067	0.915	0.820	0.053	0.767	0.613	5H	0		0.894	0.056	0.838	0.785	0.056	0.729
1.1		0.2		4h	0	1.100	0.036	1.064	0.970	0.030	0.940	0.817	4H	0	1.100	1.010	0.040	0.970	0.921	0.038	0.883
				6g	0.017	1.083	0.056	1.027	0.953	0.048	0.905	0.782									
	0.25			4h	0	1.100	0.042	1.058	0.938	0.034	0.904	0.750	4H	0	1.100	0.983	0.045	0.938	0.874	0.045	0.829
				6g	0.018	1.082	0.067	1.015	0.920	0.053	0.867	0.713	5H	0		0.994	0.056	0.938	0.885	0.056	0.829
1.2		0.2		4h	0	1.200	0.036	1.164	1.070	0.030	1.040	0.917	4H	0	1.200	1.110	0.040	1.070	1.021	0.038	0.983
				6g	0.017	1.183	0.056	1.127	1.053	0.048	1.005	0.882									
	0.25			4h	0	1.200	0.042	1.158	1.038	0.034	1.004	0.850	4H	0	1.200	1.083	0.045	1.038	0.974	0.045	0.929
				6g	0.018	1.182	0.067	1.115	1.020	0.053	0.967	0.813	5H	0		1.094	0.056	1.038	0.985	0.056	0.929
1.4		0.2		4h	0	1.400	0.036	1.364	1.270	0.030	1.240	1.117	4H	0	1.400	1.310	0.040	1.270	1.221	0.038	1.183
				6g	0.017	1.383	0.056	1.327	1.253	0.048	1.205	1.082									
	0.3			4h	0	1.400	0.048	1.352	1.205	0.036	1.169	0.984	4H	0	1.400	1.253	0.048	1.205	1.128	0.053	1.075
				6g	0.018	1.382	0.075	1.307	1.187	0.056	1.131	0.946	5H	0		1.265	0.060	1.205	1.142	0.067	1.075
													6H	0		1.280	0.075	1.205	1.160	0.085	1.075
1.6		0.2		4h	0	1.600	0.036	1.564	1.470	0.032	1.438	1.315	4H	0	1.600	1.512	0.042	1.470	1.421	0.038	1.383
				6g	0.017	1.583	0.056	1.527	1.453	0.050	1.403	1.280									
	0.35			4h	0	1.600	0.053	1.547	1.373	0.040	1.333	1.117	4H	0	1.600	1.426	0.053	1.373	1.284	0.063	1.221
				6g	0.019	1.581	0.085	1.496	1.354	0.063	1.291	1.075	5H	0		1.440	0.067	1.373	1.301	0.080	1.221
													6H	0		1.458	0.085	1.373	1.321	0.100	1.221

1	2	3	4	5	6	7	8	9	10	11	12	13	14	15	16	17	18	19	20	21	22
Nominal diameter	Pitch			External threads									Internal threads								
	Coarse	Fine	Constant	Tolerance class	Fund dev.	Major diameter			Pitch diameter			Minor diameter	Tolerance class	Fund dev.	Major diameter	Pitch diameter			Minor diameter		
						max.	tol.	min.	max.	tol.	min.	min.			min.	max.	tol.	min.	max.	tol.	min.
1.8		0.2		4h	0	1.800	0.036	1.764	1.670	0.032	1.638	1.515	4H	0	1.800	1.712	0.042	1.670	1.621	0.038	1.583
				6g	0.017	1.783	0.056	1.727	1.653	0.050	1.603	1.480									
	0.35			4h	0	1.800	0.053	1.747	1.573	0.040	1.533	1.317	4H	0	1.800	1.626	0.053	1.573	1.484	0.063	1.421
				6g	0.019	1.781	0.085	1.696	1.554	0.063	1.491	1.275	5H	0	1.800	1.640	0.067	1.573	1.501	0.080	1.421
													6H	0	1.800	1.658	0.085	1.573	1.521	0.100	1.421
2		0.25		4h	0	2.000	0.042	1.958	1.838	0.036	1.802	1.648	4H	0	2.000	1.886	0.048	1.838	1.774	0.045	1.729
				6g	0.018	1.982	0.067	1.915	1.820	0.056	1.764	1.610	5H	0	2.000	1.898	0.060	1.838	1.785	0.056	1.729
	0.4			4h	0	2.000	0.060	1.940	1.740	0.042	1.698	1.452	4H	0	2.000	1.796	0.056	1.740	1.638	0.071	1.567
				6g	0.019	1.981	0.095	1.886	1.721	0.067	1.654	1.408	5H	0	2.000	1.811	0.071	1.740	1.657	0.090	1.567
													6H	0	2.000	1.830	0.090	1.740	1.679	0.112	1.567
2.2		0.25		4h	0	2.200	0.042	2.158	2.038	0.036	2.002	1.848	4H	0	2.200	2.086	0.048	2.038	1.974	0.045	1.929
				6g	0.018	2.182	0.067	2.115	2.020	0.056	1.964	1.810	5H	0	2.200	2.098	0.060	2.038	1.985	0.056	1.929
	0.45			4h	0	2.200	0.063	2.137	1.908	0.045	1.863	1.585	4H	0	2.200	1.968	0.060	1.908	1.793	0.080	1.713
				6g	0.020	2.180	0.100	2.080	1.888	0.071	1.817	1.539	5H	0	2.200	1.983	0.075	1.908	1.813	0.100	1.713
													6H	0	2.200	2.003	0.095	1.908	1.838	0.125	1.713
2.5		0.35		4h	0	2.500	0.053	2.447	2.273	0.040	2.233	2.017	4H	0	2.500	2.326	0.053	2.273	2.184	0.063	2.121
				6g	0.019	2.481	0.085	2.396	2.254	0.063	2.191	1.975	5H	0	2.500	2.340	0.067	2.273	2.201	0.080	2.121
													6H	0	2.500	2.358	0.085	2.273	2.221	0.100	2.121
	0.45			4h	0	2.500	0.063	2.437	2.208	0.045	2.163	1.885	4H	0	2.500	2.268	0.060	2.208	2.093	0.080	2.013
				6g	0.020	2.480	0.100	2.380	2.188	0.071	2.117	1.839	5H	0	2.500	2.283	0.075	2.208	2.113	0.100	2.013
													6H	0	2.500	2.303	0.095	2.208	2.138	0.125	2.013
3		0.35		4h	0	3.000	0.053	2.947	2.773	0.042	2.731	2.515	4H	0	3.000	2.829	0.056	2.773	2.684	0.063	2.621
				6g	0.019	2.981	0.085	2.896	2.754	0.067	2.687	2.471	5H	0	3.000	2.844	0.071	2.773	2.701	0.080	2.621
													6H	0	3.000	2.863	0.090	2.773	2.721	0.100	2.621
	0.5			4h	0	3.000	0.067	2.933	2.675	0.048	2.627	2.319	5H	0	3.000	2.755	0.080	2.675	2.571	0.112	2.459
				6g	0.020	2.980	0.106	2.874	2.655	0.075	2.580	2.272	6H	0	3.000	2.775	0.100	2.675	2.599	0.140	2.459
													7H	0	3.000	2.800	0.125	2.675	2.639	0.180	2.459

Metric screw threads

1	2	3	4	5	6	7	8	9	10	11	12	13	14	15	16	17	18	19	20	21	22
Nominal diameter	Pitch		Constant	External threads									Internal threads								
	Coarse	Fine		Tolerance class	Fund dev.	Major diameter			Pitch diameter			Minor diameter	Tolerance class	Fund dev.	Major diameter	Pitch diameter			Minor diameter		
						max.	tol.	min.	max.	tol.	min.	min.			min.	max.	tol.	min.	max.	tol.	min.
3.5		0.35		4h	0	3.500	0.053	3.447	3.273	0.042	3.231	3.015	4H	0	3.500	3.329	0.056	3.273	3.184	0.063	3.121
				6g	0.019	3.481	0.085	3.396	3.254	0.067	3.187	2.971	5H	0	3.500	3.344	0.071	3.273	3.201	0.080	3.121
													6H	0	3.500	3.363	0.090	3.273	3.221	0.100	3.121
	0.6			4h	0	3.500	0.080	3.420	3.110	0.053	3.057	2.688	5H	0	3.500	3.200	0.090	3.110	2.975	0.125	2.850
				6g	0.021	3.479	0.125	3.354	3.089	0.085	3.004	2.635	6H	0	3.500	3.222	0.112	3.110	3.010	0.160	2.850
													7H	0	3.500	3.250	0.140	3.110	3.050	0.200	2.850
4		0.5		4h	0	4.000	0.067	3.933	3.675	0.048	3.627	3.319	5H	0	4.000	3.755	0.080	3.675	3.571	0.112	3.459
				6g	0.020	3.980	0.106	3.874	3.655	0.075	3.580	3.272	6H	0	4.000	3.775	0.100	3.675	3.599	0.140	3.459
													7H	0	4.000	3.800	0.125	3.675	3.639	0.180	3.459
	0.7			4h	0	4.000	0.090	3.910	3.545	0.056	3.489	3.058	5H	0	4.000	3.640	0.095	3.545	3.382	0.140	3.242
				6g	0.022	3.978	0.140	3.838	3.523	0.090	3.433	3.002	6H	0	4.000	3.663	0.118	3.545	3.422	0.180	3.242
													7H	0	4.000	3.695	0.150	3.545	3.466	0.224	3.242
4.5		0.5		4h	0	4.500	0.067	4.433	4.175	0.048	4.127	3.819	5H	0	4.500	4.255	0.080	4.175	4.071	0.112	3.959
				6g	0.020	4.480	0.106	4.374	4.155	0.075	4.080	3.772	6H	0	4.500	4.275	0.100	4.175	4.099	0.140	3.959
													7H	0	4.500	4.300	0.125	4.175	4.139	0.180	3.959
	0.75			4h	0	4.500	0.090	4.410	4.013	0.056	3.957	3.495	5H	0	4.500	4.108	0.095	4.013	3.838	0.150	3.688
				6g	0.022	4.478	0.140	4.338	3.991	0.090	3.901	3.439	6H	0	4.500	4.131	0.118	4.013	3.878	0.190	3.688
													7H	0	4.500	4.163	0.150	4.013	3.924	0.236	3.688
5		0.5		4h	0	5.000	0.067	4.933	4.675	0.048	4.627	4.319	5H	0	5.000	4.755	0.080	4.675	4.571	0.112	4.459
				6g	0.020	4.980	0.106	4.874	4.655	0.075	4.580	4.272	6H	0	5.000	4.775	0.100	4.675	4.599	0.140	4.459
													7H	0	5.000	4.800	0.125	4.675	4.639	0.180	4.459
	0.8			4h	0	5.000	0.095	4.905	4.480	0.060	4.420	3.927	5H	0	5.000	4.580	0.100	4.480	4.294	0.160	4.134
				6g	0.024	4.976	0.150	4.826	4.456	0.095	4.361	3.868	6H	0	5.000	4.605	0.125	4.480	4.334	0.200	4.134
													7H	0	5.000	4.640	0.160	4.480	4.384	0.250	4.134
5.5		0.5		4h	0	5.500	0.067	5.433	5.175	0.048	5.127	4.819	5H	0	5.500	5.255	0.080	5.175	5.071	0.112	4.959
				6g	0.020	5.480	0.106	5.374	5.155	0.075	5.080	4.772	6H	0	5.500	5.275	0.100	5.175	5.099	0.140	4.959
													7H	0	5.500	5.300	0.125	5.175	5.139	0.180	4.959

1	2	3	4	5	6	7	8	9	10	11	12	13	14	15	16	17	18	19	20	21	22
Nominal diameter	Pitch			External threads									Internal threads								
	Coarse	Fine	Constant	Tolerance class	Fund dev.	Major diameter			Pitch diameter			Minor diameter	Tolerance class	Fund dev.	Major diameter	Pitch diameter			Minor diameter		
						max.	tol.	min.	max.	tol.	min.	min.			min.	max.	tol.	min.	max.	tol.	min.
6	1	0.75		4h	0	6.000	0.090	5.910	5.513	0.063	5.450	4.988	5H	0	6.000	5.619	0.106	5.513	5.338	0.150	5.188
				6g	0.022	5.978	0.140	5.838	5.491	0.100	5.391	4.929	6H	0	6.000	5.645	0.132	5.513	5.378	0.190	5.188
													7H	0	6.000	5.683	0.170	5.513	5.424	0.236	5.188
	1			4h	0	6.000	0.112	5.888	5.350	0.071	5.279	4.663	5H	0	6.000	5.468	0.118	5.350	5.107	0.190	4.917
				6g	0.026	5.974	0.180	5.794	5.324	0.112	5.212	4.597	6H	0	6.000	5.500	0.150	5.350	5.153	0.236	4.917
				8g	0.026	5.974	0.280	5.694	5.324	0.180	5.144	4.528	7H	0	6.000	5.540	0.190	5.350	5.217	0.300	4.917
7	1	0.75		4h	0	7.000	0.090	6.910	6.513	0.063	6.450	5.988	5H	0	7.000	6.619	0.106	6.513	6.338	0.150	6.188
				6g	0.022	6.978	0.140	6.838	6.491	0.100	6.391	5.929	6H	0	7.000	6.645	0.132	6.513	6.378	0.190	6.188
													7H	0	7.000	6.683	0.170	6.513	6.424	0.236	6.188
	1			4h	0	7.000	0.112	6.888	6.350	0.071	6.279	5.663	5H	0	7.000	6.468	0.118	6.350	6.107	0.190	5.917
				6g	0.026	6.974	0.180	6.794	6.324	0.112	6.212	5.596	6H	0	7.000	6.500	0.150	6.350	6.153	0.236	5.917
				8g	0.026	6.974	0.280	6.694	6.324	0.180	6.144	5.528	7H	0	7.000	6.540	0.190	6.350	6.217	0.300	5.917
8			0.75	4h	0	8.000	0.090	7.910	7.513	0.063	7.450	6.988	5H	0	8.000	7.619	0.106	7.513	7.338	0.150	7.188
				6g	0.022	7.978	0.140	7.838	7.491	0.100	7.391	6.929	6H	0	8.000	7.645	0.132	7.513	7.378	0.190	7.188
													7H	0	8.000	7.683	0.170	7.513	7.424	0.236	7.188
		1		4h	0	8.000	0.112	7.888	7.350	0.071	7.279	6.663	5H	0	8.000	7.468	0.118	7.350	7.107	0.190	6.917
				6g	0.026	7.974	0.180	7.794	7.324	0.112	7.212	6.596	6H	0	8.000	7.500	0.150	7.350	7.153	0.236	6.917
				8g	0.026	7.974	0.280	7.694	7.324	0.180	7.144	6.528	7H	0	8.000	7.540	0.190	7.350	7.217	0.300	6.917
	1.25			4h	0	8.000	0.132	7.868	7.188	0.075	7.113	6.343	5H	0	8.000	7.313	0.125	7.188	6.859	0.212	6.647
				6g	0.028	7.972	0.212	7.760	7.160	0.118	7.042	6.272	6H	0	8.000	7.348	0.160	7.188	6.912	0.265	6.647
				8g	0.028	7.972	0.335	7.637	7.160	0.190	6.970	6.200	7H	0	8.000	7.388	0.200	7.188	6.982	0.335	6.647
9			0.75	4h	0	9.000	0.090	8.910	8.513	0.063	8.450	7.988	5H	0	9.000	8.619	0.106	8.513	8.338	0.150	8.188
				6g	0.022	8.978	0.140	8.838	8.491	0.100	8.391	7.929	6H	0	9.000	8.645	0.132	8.513	8.378	0.190	8.188
													7H	0	9.000	8.683	0.170	8.513	8.424	0.236	8.188
		1		4h	0	9.000	0.112	8.888	8.350	0.071	8.279	7.663	5H	0	9.000	8.468	0.118	8.350	8.107	0.190	7.917
				6g	0.026	8.974	0.180	8.794	8.324	0.112	8.212	7.596	6H	0	9.000	8.500	0.150	8.350	8.153	0.236	7.917
				8g	0.026	8.974	0.280	8.694	8.324	0.180	8.144	7.528	7H	0	9.000	8.540	0.190	8.350	8.217	0.300	7.917
	1.25			4h	0	9.000	0.132	8.868	8.188	0.075	8.113	7.343	5H	0	9.000	8.313	0.125	8.188	7.859	0.212	7.647
				6g	0.028	8.972	0.212	8.760	8.160	0.118	8.042	7.272	6H	0	9.000	8.348	0.160	8.188	7.912	0.265	7.647
				8g	0.028	8.972	0.335	8.637	8.160	0.190	7.970	7.200	7H	0	9.000	8.388	0.200	8.188	7.982	0.335	7.647

Metric screw threads

1	2	3	4	5	6	7	8	9	10	11	12	13	14	15	16	17	18	19	20	21	22
Nominal diameter	Pitch			External threads									Internal threads								
	Coarse	Fine	Constant	Tolerance class	Fund dev.	Major diameter			Pitch diameter			Minor diameter	Tolerance class	Fund dev.	Major diameter	Pitch diameter			Minor diameter		
						max.	tol.	min.	max.	tol.	min.	min.			min.	max.	tol.	min.	max.	tol.	min.
10			0.75	4h	0	10.000	0.090	9.910	9.513	0.063	9.450	8.988	5H	0	10.000	9.619	0.106	9.513	9.338	0.150	9.188
				6g	0.022	9.978	0.140	9.838	9.491	0.100	9.391	8.929	6H	0	10.000	9.645	0.132	9.513	9.378	0.190	9.188
													7H	0	10.000	9.683	0.170	9.513	9.424	0.236	9.188
			1	4h	0	10.000	0.112	9.888	9.350	0.071	9.279	8.663	5H	0	10.000	9.468	0.118	9.350	9.107	0.190	8.917
				6g	0.026	9.974	0.180	9.794	9.324	0.112	9.212	8.596	6H	0	10.000	9.500	0.150	9.350	9.153	0.236	8.917
				8g	0.026	9.974	0.280	9.694	9.324	0.180	9.144	8.528	7H	0	10.000	9.540	0.190	9.350	9.217	0.300	8.917
		1.25		4h	0	10.000	0.132	9.868	9.188	0.075	9.113	8.343	5H	0	10.000	9.313	0.125	9.188	8.859	0.212	8.647
				6g	0.028	9.972	0.212	9.760	9.160	0.118	9.042	8.272	6H	0	10.000	9.348	0.160	9.188	8.912	0.265	8.647
				8g	0.028	9.972	0.335	9.637	9.160	0.190	8.970	8.200	7H	0	10.000	9.388	0.200	9.188	8.982	0.335	8.647
	1.5			4h	0	10.000	0.150	9.850	9.026	0.085	8.941	8.018	5H	0	10.000	9.166	0.140	9.026	8.612	0.236	8.376
				6g	0.032	9.968	0.236	9.732	8.994	0.132	8.862	7.938	6H	0	10.000	9.206	0.180	9.026	8.676	0.300	8.376
				8g	0.032	9.968	0.375	9.593	8.994	0.212	8.782	7.858	7H	0	10.000	9.250	0.224	9.026	8.751	0.375	8.376
11			0.75	4h	0	11.000	0.090	10.910	10.513	0.063	10.450	9.988	5H	0	11.000	10.619	0.106	10.513	10.338	0.150	10.188
				6g	0.022	10.978	0.140	10.838	10.491	0.100	10.391	9.929	6H	0	11.000	10.645	0.132	10.513	10.378	0.190	10.188
													7H	0	11.000	10.683	0.170	10.513	10.424	0.236	10.188
			1	4h	0	11.000	0.112	10.888	10.350	0.071	10.279	9.663	5H	0	11.000	10.468	0.118	10.350	10.107	0.190	9.917
				6g	0.026	10.974	0.180	10.794	10.324	0.112	10.212	9.596	6H	0	11.000	10.500	0.150	10.350	10.153	0.236	9.917
				8g	0.026	10.974	0.280	10.694	10.324	0.180	10.144	9.528	7H	0	11.000	10.540	0.190	10.350	10.217	0.300	9.917
	1.5			4h	0	11.000	0.150	10.850	10.026	0.085	9.941	9.018	5H	0	11.000	10.166	0.140	10.026	9.612	0.236	9.376
				6g	0.032	10.968	0.236	10.732	9.994	0.132	9.862	8.938	6H	0	11.000	10.206	0.180	10.026	9.676	0.300	9.376
				8g	0.032	10.968	0.375	10.593	9.994	0.212	9.782	8.858	7H	0	11.000	10.250	0.224	10.026	9.751	0.375	9.376
12			1	4h	0	12.000	0.112	11.888	11.350	0.075	11.275	10.659	5H	0	12.000	11.475	0.125	11.350	11.107	0.190	10.917
				6g	0.026	11.974	0.180	11.794	11.324	0.118	11.206	10.590	6H	0	12.000	11.510	0.160	11.350	11.153	0.236	10.917
				8g	0.026	11.974	0.280	11.694	11.324	0.190	11.134	10.518	7H	0	12.000	11.550	0.200	11.350	11.217	0.300	10.917
		1.25		4h	0	12.000	0.132	11.868	11.188	0.085	11.103	10.333	5H	0	12.000	11.328	0.140	11.188	10.859	0.212	10.647
				6g	0.028	11.972	0.212	11.760	11.160	0.132	11.028	10.257	6H	0	12.000	11.368	0.180	11.188	10.912	0.265	10.647
				8g	0.028	11.972	0.335	11.637	11.160	0.212	10.948	10.177	7H	0	12.000	11.412	0.224	11.188	10.982	0.335	10.647
			1.5	4h	0	12.000	0.150	11.850	11.026	0.090	10.936	10.012	5H	0	12.000	11.176	0.150	11.026	10.612	0.236	10.376
				6g	0.032	11.968	0.236	11.732	10.994	0.140	10.854	9.930	6H	0	12.000	11.216	0.190	11.026	10.676	0.300	10.376
				8g	0.032	11.968	0.375	11.593	10.994	0.224	10.770	9.846	7H	0	12.000	11.262	0.236	11.026	10.751	0.375	10.376
	1.75			4h	0	12.000	0.170	11.830	10.863	0.095	10.768	9.692	5H	0	12.000	11.023	0.160	10.863	10.371	0.265	10.106
				6g	0.034	11.966	0.265	11.701	10.829	0.150	10.679	9.602	6H	0	12.000	11.063	0.200	10.863	10.441	0.335	10.106
				8g	0.034	11.966	0.425	11.541	10.829	0.236	10.593	9.516	7H	0	12.000	11.113	0.250	10.863	10.531	0.425	10.106

1	2	3	4	5	6	7	8	9	10	11	12	13	14	15	16	17	18	19	20	21	22
Nominal diameter	Pitch			External threads									Internal threads								
	Coarse	Fine	Constant	Tolerance class	Fund dev.	Major diameter			Pitch diameter			Minor diameter	Tolerance class	Fund dev.	Major diameter	Pitch diameter			Minor diameter		
						max.	tol.	min.	max.	tol.	min.	min.			min.	max.	tol.	min.	max.	tol.	min.
14			1	4h	0	14.000	0.112	13.888	13.350	0.075	13.275	12.659	5H	0	14.000	13.475	0.125	13.350	13.107	0.190	12.917
				6g	0.026	13.974	0.180	13.794	13.324	0.118	13.206	12.590	6H	0	14.000	13.510	0.160	13.350	13.153	0.236	12.917
				8g	0.026	13.974	0.280	13.694	13.324	0.190	13.134	12.518	7H	0	14.000	13.550	0.200	13.350	13.217	0.300	12.917
			1.25	4h	0	14.000	0.132	13.868	13.188	0.085	13.103	12.333	5H	0	14.000	13.328	0.140	13.188	12.859	0.212	12.647
				6g	0.028	13.972	0.212	13.760	13.160	0.132	13.028	12.257	6H	0	14.000	13.368	0.180	13.188	12.912	0.265	12.647
				8g	0.028	13.972	0.335	13.637	13.160	0.212	12.948	12.177	7H	0	14.000	13.412	0.224	13.188	12.982	0.335	12.647
		1.5		4h	0	14.000	0.150	13.850	13.026	0.090	12.936	12.012	5H	0	14.000	13.176	0.150	13.026	12.612	0.236	12.376
				6g	0.032	13.968	0.236	13.732	12.994	0.140	12.854	11.930	6H	0	14.000	13.216	0.190	13.026	12.676	0.300	12.376
				8g	0.032	13.968	0.375	13.593	12.994	0.224	12.770	11.846	7H	0	14.000	13.262	0.236	13.026	12.751	0.375	12.376
	2			4h	0	14.000	0.180	13.820	12.701	0.100	12.601	11.369	5H	0	14.000	12.871	0.170	12.701	12.135	0.300	11.835
				6g	0.038	13.962	0.280	13.682	12.663	0.160	12.503	11.271	6H	0	14.000	12.913	0.212	12.701	12.210	0.375	11.835
				8g	0.038	13.962	0.450	13.512	12.663	0.250	12.413	11.181	7H	0	14.000	12.966	0.265	12.701	12.310	0.475	11.835
15			1	4h	0	15.000	0.112	14.888	14.350	0.075	14.275	13.659	5H	0	15.000	14.475	0.125	14.350	14.107	0.190	13.917
				6g	0.026	14.974	0.180	14.794	14.324	0.118	14.206	13.590	6H	0	15.000	14.510	0.160	14.350	14.153	0.236	13.917
				8g	0.026	14.974	0.280	14.694	14.324	0.190	14.134	13.518	7H	0	15.000	14.550	0.200	14.350	14.217	0.300	13.917
			1.5	4h	0	15.000	0.150	14.850	14.026	0.090	13.936	13.012	5H	0	15.000	14.176	0.150	14.026	13.612	0.236	13.376
				6g	0.032	14.968	0.236	14.732	13.994	0.140	13.854	12.930	6H	0	15.000	14.216	0.190	14.026	13.676	0.300	13.376
				8g	0.032	14.968	0.375	14.593	13.994	0.224	13.770	12.846	7H	0	15.000	14.262	0.236	14.026	13.751	0.375	13.376
16			1	4h	0	16.000	0.112	15.888	15.350	0.075	15.275	14.659	5H	0	16.000	15.475	0.125	15.350	15.107	0.190	14.917
				6g	0.026	15.974	0.180	15.794	15.324	0.118	15.206	14.590	6H	0	16.000	15.510	0.160	15.350	15.153	0.236	14.917
				6g	0.026	15.974	0.280	15.694	15.324	0.190	15.134	14.518	7H	0	16.000	15.550	0.200	15.350	15.217	0.300	14.917
		1.5		4h	0	16.000	0.150	15.850	15.026	0.090	14.936	14.012	5H	0	16.000	15.176	0.150	15.026	14.612	0.236	14.376
				6g	0.032	15.968	0.236	15.732	14.994	0.140	14.854	13.930	6H	0	16.000	15.216	0.190	15.026	14.676	0.300	14.376
				8g	0.032	15.968	0.375	15.593	14.994	0.224	14.770	13.846	7H	0	16.000	15.262	0.236	15.026	14.751	0.375	14.376
	2			4h	0	16.000	0.180	15.820	14.701	0.100	14.601	13.369	5H	0	16.000	14.871	0.170	14.701	14.135	0.300	13.835
				6g	0.038	15.962	0.280	15.682	14.663	0.160	14.503	13.271	6H	0	16.000	14.913	0.212	14.701	14.210	0.375	13.835
				8g	0.038	15.962	0.450	15.512	14.663	0.250	14.413	13.181	7H	0	16.000	14.966	0.265	14.701	14.310	0.475	13.835
17			1	4h	0	17.000	0.112	16.888	16.350	0.075	16.275	15.659	5H	0	17.000	16.475	0.125	16.350	16.107	0.190	15.917
				6g	0.026	16.974	0.180	16.794	16.324	0.118	16.206	15.590	6H	0	17.000	16.510	0.160	16.350	16.153	0.236	15.917
				8g	0.026	16.974	0.280	16.694	16.324	0.190	16.134	15.518	7H	0	17.000	16.550	0.200	16.350	16.217	0.300	15.917
			1.5	4h	0	17.000	0.150	16.850	16.026	0.090	15.936	15.012	5H	0	17.000	16.176	0.150	16.026	15.612	0.236	15.376
				6g	0.032	16.968	0.236	16.732	15.994	0.140	15.854	14.930	6H	0	17.000	16.216	0.190	16.026	15.676	0.300	15.376
				8g	0.032	16.968	0.375	16.593	15.994	0.224	15.770	14.846	7H	0	17.000	16.262	0.236	16.026	15.751	0.375	15.376

Metric screw threads

1	2	3	4	5	6	7	8	9	10	11	12	13	14	15	16	17	18	19	20	21	22
Nominal diameter	Pitch			External threads									Internal threads								
	Coarse	Fine	Constant	Tolerance class	Fund dev.	Major diameter			Pitch diameter			Minor diameter	Tolerance class	Fund dev.	Major diameter	Pitch diameter			Minor diameter		
						max.	tol.	min.	max.	tol.	min.	min.			min.	max.	tol.	min.	max.	tol.	min.
18			1	4h	0	18.000	0.112	17.888	17.350	0.075	17.275	16.659	5H	0	18.000	17.475	0.125	17.350	17.107	0.190	16.917
				6g	0.026	17.974	0.180	17.794	17.324	0.118	17.206	16.590	6H	0	18.000	17.510	0.160	17.350	17.153	0.236	16.917
				8g	0.026	17.974	0.280	17.694	17.324	0.190	17.134	16.518	7H	0	18.000	17.550	0.200	17.350	17.217	0.300	16.917
		1.5		4h	0	18.000	0.150	17.850	17.026	0.090	16.936	16.012	5H	0	18.000	17.176	0.150	17.026	16.612	0.236	16.376
				6g	0.032	17.968	0.236	17.732	16.994	0.140	16.854	15.930	6H	0	18.000	17.216	0.190	17.026	16.676	0.300	16.376
				8g	0.032	17.968	0.375	17.593	16.994	0.224	16.770	15.846	7H	0	18.000	17.262	0.236	17.026	16.751	0.375	16.376
			2	4h	0	18.000	0.180	17.820	16.701	0.100	16.601	15.369	5H	0	18.000	16.871	0.170	16.701	16.135	0.300	15.835
				6g	0.038	17.962	0.280	17.682	16.663	0.160	16.503	15.271	6H	0	18.000	16.913	0.212	16.701	16.210	0.375	15.835
				8g	0.038	17.962	0.450	17.512	16.663	0.250	16.413	15.181	7H	0	18.000	16.966	0.265	16.701	16.310	0.475	15.835
	2.5			4h	0	18.000	0.212	17.788	16.376	0.106	16.270	14.730	5H	0	18.000	16.556	0.180	16.376	15.649	0.355	15.294
				6g	0.042	17.958	0.335	17.623	16.334	0.170	16.164	14.624	6H	0	18.000	16.600	0.224	16.376	15.744	0.450	15.294
				8g	0.042	17.958	0.530	17.428	16.334	0.265	16.069	14.529	7H	0	18.000	16.656	0.280	16.376	15.854	0.560	15.294
20			1	4h	0	20.000	0.112	19.888	19.350	0.075	19.275	18.659	5H	0	20.000	19.475	0.125	19.350	19.107	0.190	18.917
				6g	0.026	19.974	0.180	19.794	19.324	0.118	19.206	18.590	6H	0	20.000	19.510	0.160	19.350	19.153	0.236	18.917
				8g	0.026	19.974	0.280	19.694	19.324	0.190	19.134	18.518	7H	0	20.000	19.550	0.200	19.350	19.217	0.300	18.917
		1.5		4h	0	20.000	0.150	19.850	19.026	0.090	18.936	18.012	5H	0	20.000	19.176	0.150	19.026	18.612	0.236	18.376
				6g	0.032	19.968	0.236	19.732	18.994	0.140	18.854	17.930	6H	0	20.000	19.216	0.190	19.026	18.676	0.300	18.376
				8g	0.032	19.968	0.375	19.593	18.994	0.224	18.770	17.846	7H	0	20.000	19.262	0.236	19.026	18.751	0.375	18.376
			2	4h	0	20.000	0.180	19.820	18.701	0.100	18.601	17.369	5H	0	20.000	18.871	0.170	18.701	18.135	0.300	17.835
				6g	0.038	19.962	0.280	19.682	18.663	0.160	18.503	17.271	6H	0	20.000	18.913	0.212	18.701	18.210	0.375	17.835
				8g	0.038	19.962	0.450	19.512	18.663	0.250	18.413	17.181	7H	0	20.000	18.966	0.265	18.701	18.310	0.475	17.835
	2.5			4h	0	20.000	0.212	19.788	18.376	0.106	18.270	16.730	5H	0	20.000	18.556	0.180	18.376	17.649	0.355	17.294
				6g	0.042	19.958	0.335	19.623	18.334	0.170	18.164	16.624	6H	0	20.000	18.600	0.224	18.376	17.744	0.450	17.294
				8g	0.042	19.958	0.530	19.428	18.334	0.265	18.069	16.529	7H	0	20.000	18.656	0.280	18.376	17.854	0.560	17.294

1	2	3	4	5	6	7	8	9	10	11	12	13	14	15	16	17	18	19	20	21	22
Nominal diameter	Pitch			External threads									Internal threads								
	Coarse	Fine	Constant	Tolerance class	Fund dev.	Major diameter			Pitch diameter			Minor diameter	Tolerance class	Fund dev.	Major diameter	Pitch diameter			Minor diameter		
						max.	tol.	min.	max.	tol.	min.	min.			min.	max.	tol.	min.	max.	tol.	min.
22			1	4h	0	22.000	0.112	21.888	21.350	0.075	21.275	20.659	5H	0	22.000	21.475	0.125	21.350	21.107	0.190	20.917
				6g	0.026	21.974	0.180	21.794	21.324	0.118	21.206	20.590	6H	0	22.000	21.510	0.160	21.350	21.153	0.236	20.917
				8g	0.026	21.974	0.280	21.694	21.324	0.190	21.134	20.518	7H	0	22.000	21.550	0.200	21.350	21.217	0.300	20.917
		1.5		4h	0	22.000	0.150	21.850	21.026	0.090	20.936	20.012	5H	0	22.000	21.176	0.150	21.026	20.612	0.236	20.376
				6g	0.032	21.968	0.236	21.732	20.994	0.140	20.854	19.930	6H	0	22.000	21.216	0.190	21.026	20.676	0.300	20.376
				8g	0.032	21.968	0.375	21.593	20.994	0.224	20.770	19.846	7H	0	22.000	21.262	0.236	21.026	20.751	0.375	20.376
			2	4h	0	22.000	0.180	21.820	20.701	0.100	20.601	19.369	5H	0	22.000	20.871	0.170	20.701	20.135	0.300	19.835
				6g	0.038	21.962	0.280	21.682	20.663	0.160	20.503	19.271	6H	0	22.000	20.913	0.212	20.701	20.210	0.375	19.835
				8g	0.038	21.962	0.450	21.512	20.663	0.250	20.413	19.181	7H	0	22.000	20.966	0.265	20.701	20.310	0.475	19.835
	2.5			4h	0	22.000	0.212	21.788	20.376	0.106	20.270	18.730	5H	0	22.000	20.556	0.180	20.376	19.649	0.335	19.294
				6g	0.042	21.958	0.335	21.623	20.334	0.170	20.164	18.624	6H	0	22.000	20.600	0.224	20.376	19.744	0.450	19.294
				8g	0.042	21.958	0.530	21.428	20.334	0.265	20.069	18.529	7H	0	22.000	20.656	0.280	20.376	19.854	0.560	19.294
24			1	4h	0	24.000	0.112	23.888	23.350	0.080	23.270	22.654	5H	0	24.000	23.482	0.132	23.350	23.107	0.190	22.917
				6g	0.026	23.974	0.180	23.794	23.324	0.125	23.199	22.583	6H	0	24.000	23.520	0.170	23.350	23.153	0.236	22.917
				8g	0.026	23.974	0.280	23.694	23.324	0.200	23.124	22.508	7H	0	24.000	23.562	0.212	23.350	23.217	0.300	22.917
		1.5		4h	0	24.000	0.150	23.850	23.026	0.095	22.931	22.007	5H	0	24.000	23.186	0.160	23.026	22.612	0.236	22.376
				6g	0.032	23.968	0.236	23.732	22.994	0.150	22.844	21.920	6H	0	24.000	23.226	0.200	23.026	22.676	0.300	22.376
				8g	0.032	23.968	0.375	23.593	22.994	0.236	22.758	21.834	7H	0	24.000	23.276	0.250	23.026	22.751	0.375	22.376
			2	4h	0	24.000	0.180	23.820	22.701	0.106	22.595	21.363	5H	0	24.000	22.881	0.180	22.701	22.135	0.300	21.835
				6g	0.038	23.962	0.280	23.682	22.663	0.170	22.493	21.261	6H	0	24.000	22.925	0.224	22.701	22.210	0.375	21.835
				8g	0.038	23.962	0.450	23.512	22.663	0.265	22.398	21.166	7H	0	24.000	22.981	0.280	22.701	22.310	0.475	21.835
	3			4h	0	24.000	0.236	23.764	22.051	0.125	21.926	20.078	5H	0	24.000	22.263	0.212	22.051	21.152	0.400	20.752
				6g	0.048	23.952	0.375	23.577	22.003	0.200	21.803	19.955	6H	0	24.000	22.316	0.265	22.051	21.252	0.500	20.752
				8g	0.048	23.952	0.600	23.352	22.003	0.315	21.688	19.840	7H	0	24.000	22.386	0.335	22.051	21.382	0.630	20.752
25			1	4h	0	25.000	0.112	24.888	24.350	0.080	24.270	23.654	5H	0	25.000	24.482	0.132	24.350	24.107	0.190	23.917
				6g	0.026	24.974	0.180	24.794	24.324	0.125	24.199	23.583	6H	0	25.000	24.520	0.170	24.350	24.153	0.236	23.917
				8g	0.026	24.974	0.280	24.694	24.324	0.200	24.124	23.508	7H	0	25.000	24.562	0.212	24.350	24.217	0.300	23.917
		1.5		4h	0	25.000	0.150	24.850	24.026	0.095	23.931	23.007	5H	0	25.000	24.186	0.160	24.026	23.612	0.236	23.376
				6g	0.032	24.968	0.236	24.732	23.994	0.150	23.844	22.920	6H	0	25.000	24.226	0.200	24.026	23.676	0.300	23.376
				8g	0.032	24.968	0.375	24.593	23.994	0.236	23.758	22.834	7H	0	25.000	24.276	0.250	24.026	23.751	0.375	23.376
			2	4h	0	25.000	0.180	24.820	23.701	0.106	23.595	22.363	5H	0	25.000	23.881	0.180	23.701	23.135	0.300	22.835
				6g	0.038	24.962	0.280	24.682	23.663	0.170	23.493	22.261	6H	0	25.000	23.925	0.224	23.701	23.210	0.375	22.835
				8g	0.038	24.962	0.450	24.512	23.663	0.265	23.398	22.166	7H	0	25.000	23.981	0.280	23.701	23.310	0.475	22.835

Metric screw threads

1	2	3	4	5	6	7	8	9	10	11	12	13	14	15	16	17	18	19	20	21	22
Nominal diameter	Pitch			External threads									Internal threads								
	Coarse	Fine	Constant	Tolerance class	Fund dev.	Major diameter			Pitch diameter			Minor diameter	Tolerance class	Fund dev.	Major diameter	Pitch diameter			Minor diameter		
						max.	tol.	min.	max.	tol.	min.	min.			min.	max.	tol.	min.	max.	tol.	min.
26			1.5	4h	0	26.000	0.150	25.850	25.026	0.095	24.931	24.007	5H	0	26.000	25.186	0.160	25.026	24.612	0.236	24.376
				6g	0.032	25.968	0.236	25.732	24.994	0.150	24.844	23.920	6H	0	26.000	25.226	0.200	25.026	24.676	0.300	24.376
				8g	0.032	25.968	0.375	25.593	24.994	0.236	24.758	23.834	7H	0	26.000	25.276	0.250	25.026	24.751	0.375	24.376
27			1	4h	0	27.000	0.112	26.888	26.350	0.080	26.270	25.654	5H	0	27.000	26.482	0.132	26.350	26.107	0.190	25.917
				6g	0.026	26.974	0.180	26.794	26.324	0.125	26.199	25.583	6H	0	27.000	26.520	0.170	26.350	26.153	0.236	25.917
				8g	0.026	26.974	0.280	26.694	26.324	0.200	26.124	25.508	7H	0	27.000	26.562	0.212	26.350	26.217	0.300	25.917
			1.5	4h	0	27.000	0.150	26.850	26.026	0.095	25.931	25.007	5H	0	27.000	26.186	0.160	26.026	25.612	0.236	25.376
				6g	0.032	26.968	0.236	26.732	25.994	0.150	25.844	24.920	6H	0	27.000	26.226	0.200	26.026	25.676	0.300	25.376
				8g	0.032	26.968	0.375	26.593	25.994	0.236	25.758	24.834	7H	0	27.000	26.276	0.250	26.026	25.751	0.375	25.376
		2		4h	0	27.000	0.180	26.820	25.701	0.106	25.595	24.363	5H	0	27.000	25.881	0.180	25.701	25.135	0.300	24.835
				6g	0.038	26.962	0.280	26.682	25.663	0.170	25.493	24.261	6H	0	27.000	25.925	0.224	25.701	25.210	0.375	24.835
				8g	0.038	26.962	0.450	26.512	25.663	0.265	25.398	24.166	7H	0	27.000	25.981	0.280	25.701	25.310	0.475	24.835
	3			4h	0	27.000	0.236	26.764	25.051	0.125	24.926	23.078	5H	0	27.000	25.263	0.212	25.051	24.152	0.400	23.752
				6g	0.048	26.952	0.375	26.577	25.003	0.200	24.803	22.955	6H	0	27.000	25.316	0.265	25.051	24.252	0.500	23.752
				8g	0.048	26.952	0.600	26.352	25.003	0.315	24.688	22.840	7H	0	27.000	25.386	0.335	25.051	24.382	0.630	23.752
28			1	4h	0	28.000	0.112	27.888	27.350	0.080	27.270	26.654	5H	0	28.000	27.482	0.132	27.350	27.107	0.190	26.917
				6g	0.026	27.974	0.180	27.794	27.324	0.125	27.199	26.583	6H	0	28.000	27.520	0.170	27.350	27.153	0.236	26.917
				8g	0.026	27.974	0.280	27.694	27.324	0.200	27.124	26.508	7H	0	28.000	27.562	0.212	27.350	27.217	0.300	26.917
			1.5	4h	0	28.000	0.150	27.850	27.026	0.095	26.931	26.007	5H	0	28.000	27.186	0.160	27.026	26.612	0.236	26.376
				6g	0.032	27.968	0.236	27.732	26.994	0.150	26.844	25.920	6H	0	28.000	27.226	0.200	27.026	26.676	0.300	26.376
				8g	0.032	27.968	0.375	27.593	26.994	0.236	26.758	25.834	7H	0	28.000	27.276	0.250	27.026	26.751	0.375	26.376
			2	4h	0	28.000	0.180	27.820	26.701	0.106	26.595	25.363	5H	0	28.000	26.881	0.180	26.701	26.135	0.300	25.835
				6g	0.038	27.962	0.280	27.682	26.663	0.170	26.493	25.261	6H	0	28.000	26.925	0.224	26.701	26.210	0.375	25.835
				8g	0.038	27.962	0.450	27.512	26.663	0.265	26.390	25.166	7H	0	28.000	26.981	0.280	26.701	26.310	0.475	25.835

1	2	3	4	5	6	7	8	9	10	11	12	13	14	15	16	17	18	19	20	21	22
Nominal diameter	Pitch			External threads									Internal threads								
	Coarse	Fine	Constant	Tolerance class	Fund dev.	Major diameter			Pitch diameter			Minor diameter	Tolerance class	Fund dev.	Major diameter	Pitch diameter			Minor diameter		
						max.	tol.	min.	max.	tol.	min.	min.			min.	max.	tol.	min.	max.	tol.	min.
30			1	4h	0	30.000	0.112	29.888	29.350	0.080	29.270	28.654	5H	0	30.000	29.482	0.132	29.350	29.107	0.190	28.917
				6g	0.026	29.974	0.180	29.794	29.324	0.125	29.199	28.583	6H	0	30.000	29.520	0.170	29.350	29.153	0.236	28.917
				8g	0.026	29.974	0.280	29.694	29.324	0.200	29.124	28.508	7H	0	30.000	29.562	0.212	29.350	29.217	0.300	28.917
			1.5	4h	0	30.000	0.150	29.850	29.026	0.095	28.931	28.007	5H	0	30.000	29.186	0.160	29.026	28.612	0.236	28.376
				6g	0.032	29.968	0.236	29.732	28.994	0.150	28.844	27.920	6H	0	30.000	29.226	0.200	29.026	28.676	0.300	28.376
				8g	0.032	29.968	0.375	29.593	28.994	0.236	28.758	27.834	7H	0	30.000	29.276	0.250	29.026	28.751	0.375	28.376
		2		4h	0	30.000	0.180	29.820	28.701	0.106	28.595	27.363	5H	0	30.000	28.881	0.180	28.701	28.135	0.300	27.835
				6g	0.038	29.962	0.280	29.682	28.663	0.170	28.493	27.261	6H	0	30.000	28.925	0.224	28.701	28.210	0.375	27.835
				8g	0.038	29.962	0.450	29.512	28.663	0.265	28.398	27.166	7H	0	30.000	28.981	0.280	28.701	28.310	0.475	27.835
			3	4h	0	30.000	0.236	29.764	28.051	0.125	27.926	26.078	5H	0	30.000	28.263	0.212	28.051	27.152	0.400	26.752
				6g	0.048	29.952	0.375	29.577	28.003	0.200	27.803	25.955	6H	0	30.000	28.316	0.265	28.051	27.252	0.500	26.752
				8g	0.048	29.952	0.600	29.352	28.003	0.315	27.688	25.840	7H	0	30.000	28.386	0.335	28.051	27.382	0.630	26.752
	3.5			4h	0	30.000	0.265	29.735	27.727	0.132	27.595	25.438	5H	0	30.000	27.951	0.224	27.727	26.661	0.450	26.211
				6g	0.053	29.947	0.425	29.522	27.674	0.212	27.462	25.305	6H	0	30.000	28.007	0.280	27.727	26.771	0.560	26.211
				8g	0.053	29.947	0.670	29.277	27.674	0.335	27.339	25.182	7H	0	30.000	28.082	0.355	27.727	26.921	0.710	26.211
32			1.5	4h	0	32.000	0.150	31.850	31.026	0.095	30.931	30.007	5H	0	32.000	31.186	0.160	31.026	30.612	0.236	30.376
				6g	0.032	31.968	0.236	31.732	30.994	0.150	30.844	29.920	6H	0	32.000	31.226	0.200	31.026	30.676	0.300	30.376
				8g	0.032	31.968	0.375	31.593	30.994	0.236	30.758	29.834	7H	0	32.000	31.276	0.250	31.026	30.751	0.375	30.376
		2		4h	0	32.000	0.180	31.820	30.701	0.106	30.595	29.363	5H	0	32.000	30.881	0.180	30.701	30.135	0.300	29.835
				6g	0.038	31.962	0.280	31.682	30.663	0.170	30.493	29.261	6H	0	32.000	30.925	0.224	30.701	30.210	0.375	29.835
				8g	0.038	31.962	0.450	31.512	30.663	0.265	30.398	29.166	7H	0	32.000	30.981	0.280	30.701	30.310	0.475	29.835
33			1.5	4h	0	33.000	0.150	32.850	32.026	0.095	31.931	31.007	5H	0	33.000	32.186	0.160	32.026	31.612	0.236	31.376
				6g	0.032	32.968	0.236	32.732	31.994	0.150	31.844	30.920	6H	0	33.000	32.226	0.200	32.026	31.676	0.300	31.376
				8g	0.032	32.968	0.375	32.593	31.994	0.236	31.758	30.834	7H	0	33.000	32.276	0.250	32.026	31.751	0.375	31.376
		2		4h	0	33.000	0.180	32.820	31.701	0.106	31.595	30.363	5H	0	33.000	31.881	0.180	31.701	31.135	0.300	30.835
				6g	0.038	32.962	0.280	32.682	31.663	0.170	31.493	30.261	6H	0	33.000	31.925	0.224	31.701	31.210	0.375	30.835
				8g	0.038	32.962	0.450	32.512	31.663	0.265	31.398	30.166	7H	0	33.000	31.981	0.280	31.701	31.310	0.475	30.835
			3	4h	0	33.000	0.236	32.764	31.051	0.125	30.926	29.078	5H	0	33.000	31.263	0.212	31.051	31.152	0.400	29.752
				6g	0.048	32.952	0.375	32.577	31.003	0.200	30.803	28.955	6H	0	33.000	31.316	0.265	31.051	30.252	0.500	29.752
				8g	0.048	32.952	0.600	32.352	31.003	0.315	30.688	28.840	7H	0	33.000	31.386	0.335	31.051	30.382	0.630	29.752
	3.5			4h	0	33.000	0.265	32.735	30.727	0.132	30.595	28.438	5H	0	33.000	30.951	0.224	30.727	29.661	0.450	29.211
				6g	0.053	32.947	0.425	32.522	30.674	0.212	30.462	28.305	6H	0	33.000	31.007	0.280	30.727	29.771	0.560	29.211
				8g	0.053	32.947	0.670	32.277	30.674	0.335	30.339	28.182	7H	0	33.000	31.082	0.355	30.727	29.921	0.710	29.211

Metric screw threads

1	2	3	4	5	6	7	8	9	10	11	12	13	14	15	16	17	18	19	20	21	22
Nominal diameter	Pitch			External threads									Internal threads								
	Coarse	Fine	Constant	Tolerance class	Fund dev.	Major diameter			Pitch diameter			Minor diameter	Tolerance class	Fund dev.	Major diameter	Pitch diameter			Minor diameter		
						max.	tol.	min.	max.	tol.	min.	min.			min.	max.	tol.	min.	max.	tol.	min.
35			1.5	4h	0	35.000	0.150	34.850	34.026	0.095	33.931	33.007	5H	0	35.000	34.186	0.160	34.026	33.612	0.236	33.376
				6g	0.032	34.968	0.236	34.732	33.994	0.150	33.844	32.920	6H	0	35.000	34.226	0.200	34.026	33.676	0.300	33.376
				8g	0.032	34.968	0.375	34.593	33.994	0.236	33.758	32.834	7H	0	35.000	34.276	0.250	34.026	33.751	0.375	33.376
36			1.5	4h	0	36.000	0.150	35.850	35.026	0.095	34.931	34.007	5H	0	36.000	35.186	0.160	35.026	34.612	0.236	34.376
				6g	0.032	35.968	0.236	35.732	34.994	0.150	34.844	33.920	6H	0	36.000	35.226	0.200	35.026	34.676	0.300	34.376
				8g	0.032	35.968	0.375	35.593	34.994	0.236	34.758	33.834	7H	0	36.000	35.276	0.250	35.026	34.751	0.375	34.376
			2	4h	0	36.000	0.180	35.820	34.701	0.106	34.595	33.363	5H	0	36.000	34.881	0.180	34.701	34.135	0.300	33.835
				6g	0.038	35.962	0.280	35.682	34.663	0.170	34.493	33.261	6H	0	36.000	34.925	0.224	34.701	34.210	0.375	33.835
				8g	0.038	35.962	0.450	35.512	34.663	0.265	34.398	33.166	7H	0	36.000	34.981	0.280	34.701	34.310	0.475	33.835
			3	4h	0	36.000	0.236	35.764	34.051	0.125	33.926	32.078	5H	0	36.000	34.263	0.212	34.051	33.152	0.400	32.752
				6g	0.048	35.952	0.375	35.577	34.003	0.200	33.803	31.955	6H	0	36.000	34.316	0.265	34.051	33.252	0.500	32.752
				8g	0.048	35.952	0.600	35.352	34.003	0.315	33.688	31.840	7H	0	36.000	34.386	0.335	34.051	33.382	0.630	32.752
	4			4h	0	36.000	0.300	35.700	33.402	0.140	33.262	30.798	5H	0	36.000	33.638	0.236	33.402	32.145	0.475	31.670
				6g	0.060	35.940	0.475	35.465	33.342	0.224	33.118	30.654	6H	0	36.000	33.702	0.300	33.402	32.270	0.600	31.670
				8g	0.060	35.940	0.750	35.190	33.342	0.355	32.987	30.523	7H	0	36.000	33.777	0.375	33.402	32.420	0.750	31.670
38			1.5	4h	0	38.000	0.150	37.850	37.026	0.095	36.931	36.007	5H	0	38.000	37.186	0.160	37.026	37.612	0.236	36.376
				6g	0.032	37.968	0.236	37.732	36.994	0.150	36.844	35.920	6H	0	38.000	37.226	0.200	37.026	36.676	0.300	36.376
				8g	0.032	37.968	0.376	37.593	36.994	0.236	36.758	35.834	7H	0	38.000	37.276	0.250	37.026	36.751	0.375	36.376
39			1.5	4h	0	39.000	0.150	38.850	38.026	0.095	37.931	37.007	5H	0	39.000	38.186	0.160	38.026	37.612	0.236	37.376
				6g	0.032	38.968	0.263	38.732	37.994	0.150	37.844	36.920	6H	0	39.000	38.226	0.200	38.026	37.676	0.300	37.376
				8g	0.032	38.968	0.375	37.593	37.994	0.236	37.758	36.834	7H	0	39.000	38.276	0.250	38.026	37.751	0.375	37.376
			2	4h	0	39.000	0.180	38.820	37.701	0.106	37.595	36.363	5H	0	39.000	37.881	0.180	37.701	37.135	0.300	36.835
				6g	0.038	38.962	0.280	38.682	37.663	0.170	37.493	36.261	6H	0	39.000	37.925	0.224	37.701	37.210	0.375	36.835
				8g	0.038	38.962	0.450	38.512	37.663	0.265	37.398	36.166	7H	0	39.000	37.981	0.280	37.701	37.310	0.475	36.835
			3	4h	0	39.000	0.236	38.764	37.051	0.125	36.926	35.078	5H	0	39.000	37.263	0.212	37.051	36.152	0.400	35.752
				6g	0.048	38.952	0.375	38.577	37.003	0.200	36.803	34.955	6H	0	39.000	37.316	0.265	37.051	36.252	0.500	35.752
				8g	0.048	38.952	0.600	38.352	37.003	0.315	36.688	34.840	7H	0	39.000	37.386	0.335	37.051	36.382	0.630	35.752
	4			4h	0	39.000	0.300	38.700	36.402	0.140	36.262	33.798	5H	0	39.000	36.638	0.236	36.402	35.145	0.475	34.670
				6g	0.060	38.940	0.475	38.465	36.342	0.224	36.118	33.654	6H	0	39.000	36.702	0.300	36.402	35.270	0.600	34.670
				8g	0.060	38.940	0.750	38.190	36.342	0.355	35.987	33.523	7H	0	39.000	36.777	0.375	36.402	35.420	0.750	34.670

1	2	3	4	5	6	7	8	9	10	11	12	13	14	15	16	17	18	19	20	21	22
Nominal diameter	Pitch			External threads									Internal threads								
	Coarse	Fine	Constant	Tolerance class	Fund dev.	Major diameter			Pitch diameter			Minor diameter	Tolerance class	Fund dev.	Major diameter	Pitch diameter			Minor diameter		
						max.	tol.	min.	max.	tol.	min.	min.			min.	max.	tol.	min.	max.	tol.	min.
40			1.5	4h	0	40.000	0.150	39.850	39.026	0.095	38.931	38.007	5H	0	40.000	39.186	0.160	39.026	38.612	0.236	38.376
				6g	0.032	39.968	0.236	39.732	38.994	0.150	38.844	37.920	6H	0	40.000	39.226	0.200	39.026	38.676	0.300	38.376
				8g	0.032	39.968	0.375	39.593	38.994	0.236	38.758	37.834	7H	0	40.000	39.276	0.250	39.026	38.751	0.375	38.376
			2	4h	0	40.000	0.180	39.820	38.701	0.106	38.595	37.363	5H	0	40.000	38.881	0.180	38.701	38.135	0.300	37.835
				6g	0.038	39.962	0.280	39.682	38.663	0.170	38.493	37.261	6H	0	40.000	38.925	0.224	38.701	38.210	0.375	37.835
				8g	0.038	39.962	0.450	39.512	38.663	0.265	38.398	37.166	7H	0	40.000	38.981	0.280	38.701	38.310	0.475	37.835
			3	4h	0	40.000	0.236	39.764	38.051	0.125	37.926	36.078	5H	0	40.000	38.263	0.212	38.051	37.152	0.400	36.752
				6g	0.048	39.952	0.375	39.577	38.003	0.200	37.803	35.955	6H	0	40.000	38.316	0.265	38.051	37.252	0.500	36.752
				8g	0.048	39.952	0.600	39.352	38.003	0.315	37.688	35.840	7H	0	40.000	38.386	0.335	38.051	37.382	0.630	36.752
42			1.5	4h	0	42.000	0.150	41.850	41.026	0.095	40.931	40.007	5H	0	42.000	41.186	0.160	41.026	40.612	0.236	40.376
				6g	0.032	41.968	0.236	41.732	40.994	0.150	40.844	39.920	6H	0	42.000	41.226	0.200	41.026	40.676	0.300	40.376
				8g	0.032	41.968	0.375	41.593	40.994	0.236	40.758	39.834	7H	0	42.000	41.276	0.250	41.026	40.751	0.375	40.376
			2	4h	0	42.000	0.180	41.820	40.701	0.106	40.595	39.363	5H	0	42.000	40.881	0.180	40.701	40.135	0.300	39.835
				6g	0.038	41.962	0.280	41.682	40.663	0.170	40.493	39.261	6H	0	42.000	40.925	0.224	40.701	40.210	0.375	39.835
				8g	0.038	41.962	0.450	41.512	40.663	0.265	40.398	39.166	7H	0	42.000	40.981	0.280	40.701	40.310	0.475	39.835
			3	4h	0	42.000	0.236	41.764	40.051	0.125	39.926	38.078	5H	0	42.000	40.263	0.212	40.051	39.152	0.400	38.752
				6g	0.048	41.952	0.375	41.577	40.003	0.200	39.803	37.955	6H	0	42.000	40.316	0.265	40.051	39.252	0.500	38.752
				8g	0.048	41.952	0.600	41.352	40.003	0.315	39.688	37.840	7H	0	42.000	40.386	0.335	40.051	39.382	0.630	38.752
			4	4h	0	42.000	0.300	41.700	39.402	0.140	39.262	36.798	5H	0	42.000	39.638	0.236	39.402	38.145	0.475	37.670
				6g	0.060	41.940	0.475	41.465	39.342	0.224	39.118	36.654	6H	0	42.000	39.702	0.300	39.402	38.270	0.600	37.670
				8g	0.060	41.940	0.750	41.190	39.342	0.355	38.987	36.523	7H	0	42.000	39.777	0.375	39.402	38.420	0.750	37.670
	4.5			4h	0	42.000	0.315	41.685	39.077	0.150	38.927	36.155	5H	0	42.000	39.327	0.250	39.077	37.659	0.530	37.129
				6g	0.063	41.937	0.500	41.437	39.014	0.236	38.778	36.006	6H	0	42.000	39.392	0.315	39.077	37.799	0.670	37.129
				8g	0.063	41.937	0.800	41.137	39.014	0.375	38.639	35.867	7H	0	42.000	39.477	0.400	39.077	37.979	0.850	37.129

Metric screw threads

1	2	3	4	5	6	7	8	9	10	11	12	13	14	15	16	17	18	19	20	21	22
Nominal diameter	Pitch			External threads									Internal threads								
	Coarse	Fine	Constant	Tolerance class	Fund dev.	Major diameter			Pitch diameter			Minor diameter	Tolerance class	Fund dev.	Major diameter	Pitch diameter			Minor diameter		
						max.	tol.	min.	max.	tol.	min.	min.			min.	max.	tol.	min.	max.	tol.	min.
45			1.5	4h	0	45.000	0.150	44.850	44.026	0.095	43.931	43.007	5H	0	45.000	44.186	0.160	44.026	43.612	0.236	43.376
				6g	0.032	44.968	0.236	44.732	43.994	0.150	43.844	42.920	6H	0	45.000	44.226	0.200	44.026	43.676	0.300	43.376
				8g	0.032	44.968	0.375	44.593	43.994	0.236	43.758	42.834	7H	0	45.000	44.276	0.250	44.026	43.751	0.375	43.376
			2.0	4h	0	45.000	0.180	44.820	43.701	0.106	43.595	42.363	5H	0	45.000	43.881	0.180	43.701	43.135	0.300	42.835
				6g	0.038	44.962	0.280	44.682	43.663	0.170	43.493	42.261	6H	0	45.000	43.925	0.224	43.701	43.210	0.375	42.835
				8g	0.038	44.962	0.450	44.512	43.663	0.265	43.398	42.166	7H	0	45.000	43.981	0.280	43.701	43.310	0.475	42.835
			3.0	4h	0	45.000	0.236	44.764	43.051	0.125	42.926	41.078	5H	0	45.000	43.263	0.212	43.051	42.152	0.400	41.752
				6g	0.048	44.952	0.375	44.577	43.003	0.200	42.803	40.955	6H	0	45.000	43.316	0.265	43.051	42.252	0.500	41.752
				8g	0.048	44.952	0.600	44.352	43.003	0.315	42.688	40.840	7H	0	45.000	43.386	0.335	43.051	42.382	0.630	41.752
			4.0	4h	0	45.000	0.300	44.700	42.402	0.140	42.262	39.798	5H	0	45.000	42.638	0.236	42.402	41.145	0.475	40.670
				6g	0.060	44.940	0.475	44.465	42.342	0.224	42.118	39.654	6H	0	45.000	42.702	0.300	42.402	41.270	0.600	40.670
				8g	0.060	44.940	0.750	44.190	42.342	0.355	41.987	39.523	7H	0	45.000	42.777	0.375	42.402	41.420	0.750	40.670
	4.5			4h	0	45.000	0.315	44.685	42.077	0.150	41.927	39.155	5H	0	45.000	42.327	0.250	42.077	40.659	0.530	40.129
				6g	0.063	44.937	0.500	44.437	42.014	0.236	41.778	39.006	6H	0	45.000	42.392	0.315	42.077	40.799	0.670	40.129
				8g	0.063	44.937	0.800	44.137	42.014	0.375	41.639	38.867	7H	0	45.000	42.477	0.400	42.077	40.979	0.850	40.129
48			1.5	4h	0	48.000	0.150	47.850	47.026	0.100	46.926	46.002	5H	0	48.000	47.196	0.170	47.026	46.612	0.236	46.376
				6g	0.032	47.968	0.236	47.732	46.994	0.160	46.834	45.910	6H	0	48.000	47.238	0.212	47.026	46.676	0.300	46.376
				8g	0.032	47.968	0.375	47.593	46.994	0.250	46.744	45.820	7H	0	48.000	47.291	0.265	47.026	46.751	0.375	46.376
			2.0	4h	0	48.000	0.180	47.820	46.701	0.112	46.589	45.357	5H	0	48.000	46.891	0.190	46.701	46.135	0.300	45.835
				6g	0.038	47.962	0.280	47.682	46.663	0.180	46.483	45.251	6H	0	48.000	46.937	0.236	46.701	46.210	0.375	45.835
				8g	0.038	47.962	0.450	47.512	46.663	0.280	46.383	45.151	7H	0	48.000	47.001	0.300	46.701	46.310	0.475	45.835
			3.0	4h	0	48.000	0.236	47.764	46.051	0.132	45.919	44.071	5H	0	48.000	46.275	0.224	46.051	45.152	0.400	44.752
				6g	0.048	47.952	0.375	47.577	46.003	0.212	45.791	43.943	6H	0	48.000	46.331	0.280	46.051	45.252	0.500	44.752
				8g	0.048	47.952	0.600	47.352	46.003	0.335	45.668	43.820	7H	0	48.000	46.406	0.355	46.051	45.382	0.630	44.752
			4.0	4h	0	48.000	0.300	47.700	45.402	0.150	45.252	42.788	5H	0	48.000	45.652	0.250	45.402	44.145	0.475	43.670
				6g	0.060	47.940	0.475	47.465	45.342	0.236	45.106	42.642	6H	0	48.000	45.717	0.315	45.402	44.270	0.600	43.670
				8g	0.060	47.940	0.750	47.190	45.342	0.375	44.967	42.503	7H	0	48.000	45.802	0.400	45.402	44.420	0.750	43.670
	5.0			4h	0	48.000	0.335	47.665	44.752	0.160	44.592	41.512	5H	0	48.000	45.017	0.265	44.752	43.147	0.560	42.587
				6g	0.071	47.929	0.530	47.399	44.681	0.250	44.431	41.351	6H	0	48.000	45.087	0.335	44.752	43.297	0.710	42.587
				8g	0.071	47.929	0.850	47.079	44.681	0.400	44.281	41.201	7H	0	48.000	45.177	0.425	44.752	43.487	0.900	42.587

1	2	3	4	5	6	7	8	9	10	11	12	13	14	15	16	17	18	19	20	21	22
Nominal diameter	Pitch			External threads									Internal threads								
	Coarse	Fine	Constant	Tolerance class	Fund dev.	Major diameter			Pitch diameter			Minor diameter	Tolerance class	Fund dev.	Major diameter	Pitch diameter			Minor diameter		
						max.	tol.	min.	max.	tol.	min.	min.			min.	max.	tol.	min.	max.	tol.	min.
50			1.5	4h	0	50.000	0.150	49.850	49.026	0.100	48.926	48.002	5H	0	50.000	49.196	0.170	49.026	48.612	0.236	48.376
				6g	0.032	49.968	0.236	49.732	48.994	0.160	48.834	47.910	6H	0	50.000	49.238	0.212	49.026	48.676	0.300	48.376
				8g	0.032	49.968	0.375	49.593	48.994	0.250	48.744	47.820	7H	0	50.000	49.291	0.265	49.026	48.751	0.375	48.376
			2.0	4h	0	50.000	0.180	49.820	48.701	0.112	48.589	47.357	5H	0	50.000	48.891	0.190	48.701	48.135	0.300	47.835
				6g	0.038	49.962	0.280	49.682	48.663	0.180	48.483	47.251	6H	0	50.000	48.937	0.236	48.701	48.210	0.375	47.835
				8g	0.038	49.962	0.450	49.512	48.663	0.280	48.383	47.151	7H	0	50.000	49.001	0.300	48.701	48.310	0.475	47.835
			3.0	4h	0	50.000	0.236	49.764	48.051	0.132	47.919	46.071	5H	0	50.000	48.275	0.224	48.051	47.152	0.400	46.752
				6g	0.048	49.952	0.375	49.577	48.003	0.212	47.791	45.943	6H	0	50.000	48.331	0.280	48.051	47.252	0.500	46.752
				8g	0.048	49.952	0.600	49.352	48.003	0.335	47.668	45.820	7H	0	50.000	48.406	0.355	48.051	47.382	0.630	46.752
52			1.5	4h	0	52.000	0.150	51.850	51.026	0.100	50.926	50.002	5H	0	52.000	51.196	0.170	51.026	50.612	0.236	50.376
				6g	0.032	51.968	0.236	51.732	50.994	0.160	50.834	49.910	6H	0	52.000	51.238	0.212	51.026	50.676	0.300	50.376
				8g	0.032	51.968	0.375	51.593	50.994	0.250	50.744	49.820	7H	0	52.000	51.291	0.265	51.026	50.751	0.375	50.376
			2.0	4h	0	52.000	0.180	51.820	50.701	0.112	50.589	49.357	5H	0	52.000	50.891	0.190	50.701	50.135	0.300	49.835
				6g	0.038	51.962	0.280	51.682	50.663	0.180	50.483	49.251	6H	0	52.000	50.937	0.236	50.701	50.210	0.375	49.835
				8g	0.038	51.962	0.450	51.512	50.663	0.280	50.383	49.151	7H	0	52.000	51.001	0.300	50.701	50.310	0.475	49.835
			3.0	4h	0	52.000	0.236	51.764	50.051	0.132	49.919	48.071	5H	0	52.000	50.275	0.224	50.051	49.152	0.400	48.752
				6g	0.048	51.952	0.375	51.577	50.003	0.212	49.791	47.943	6H	0	52.000	50.331	0.280	50.051	49.252	0.500	48.752
				8g	0.048	51.952	0.600	51.352	50.003	0.335	49.668	47.820	7H	0	52.000	50.406	0.355	50.051	49.382	0.630	48.752
			4.0	4h	0	52.000	0.300	51.700	49.402	0.150	49.252	46.788	5H	0	52.000	49.652	0.250	49.402	48.145	0.475	47.670
				6g	0.060	51.940	0.475	51.465	49.342	0.236	49.106	46.642	6H	0	52.000	49.717	0.315	49.402	48.270	0.600	47.670
				8g	0.060	51.940	0.750	51.190	49.342	0.375	48.967	46.503	7H	0	52.000	49.802	0.400	49.402	48.420	0.750	47.670
	5.0			4h	0	52.000	0.335	51.665	48.752	0.160	48.592	45.512	5H	0	52.000	49.017	0.265	48.752	47.147	0.560	46.587
				6g	0.071	51.929	0.530	51.399	48.681	0.250	48.431	45.351	6H	0	52.000	49.087	0.335	48.752	47.297	0.710	46.587
				8g	0.071	51.929	0.850	51.079	48.681	0.400	48.281	45.201	7H	0	52.000	49.177	0.425	48.752	47.687	0.900	46.587

Metric screw threads

1	2	3	4	5	6	7	8	9	10	11	12	13	14	15	16	17	18	19	20	21	22
Nominal diameter	Pitch			External threads									Internal threads								
	Coarse	Fine	Constant	Tolerance class	Fund dev.	Major diameter			Pitch diameter			Minor diameter	Tolerance class	Fund dev.	Major diameter	Pitch diameter			Minor diameter		
						max.	tol.	min.	max.	tol.	min.	min.			min.	max.	tol.	min.	max.	tol.	min.
55			1.5	4h	0	55.000	0.150	54.850	54.026	0.100	53.926	53.002	5H	0	55.000	54.196	0.170	54.026	53.612	0.236	53.376
				6g	0.032	54.968	0.236	54.732	53.994	0.160	53.834	52.910	6H	0	55.000	54.238	0.212	54.026	53.676	0.300	53.376
				8g	0.032	54.968	0.375	54.593	53.994	0.250	53.744	52.820	7H	0	55.000	54.291	0.265	54.026	53.751	0.375	53.376
			2.0	4h	0	55.000	0.180	54.820	53.701	0.112	53.589	52.357	5H	0	55.000	53.891	0.190	53.701	53.135	0.300	52.835
				6g	0.038	54.962	0.280	54.682	53.663	0.180	53.483	52.251	6H	0	55.000	53.937	0.236	53.701	53.210	0.375	52.835
				8g	0.038	54.962	0.450	54.512	53.663	0.280	53.383	52.151	7H	0	55.000	54.001	0.300	53.701	53.310	0.475	52.835
			3.0	4h	0	55.000	0.236	54.764	53.051	0.132	52.919	51.071	5H	0	55.000	53.275	0.224	53.051	52.152	0.400	51.752
				6g	0.048	54.952	0.375	54.577	53.003	0.212	52.791	50.943	6H	0	55.000	53.331	0.280	53.051	52.252	0.500	51.752
				8g	0.048	54.952	0.600	54.352	53.003	0.335	52.668	50.820	7H	0	55.000	53.406	0.355	53.051	52.382	0.630	51.752
			4.0	4h	0	55.000	0.300	54.700	52.402	0.150	52.252	49.788	5H	0	55.000	52.652	0.250	52.402	51.145	0.475	50.670
				6g	0.060	54.940	0.475	54.465	52.342	0.236	52.106	49.642	6H	0	55.000	52.717	0.315	52.402	51.270	0.600	50.670
				8g	0.060	54.940	0.750	54.190	52.342	0.375	51.967	49.503	7H	0	55.000	52.802	0.400	52.402	51.420	0.750	50.670
56			1.5	4h	0	56.000	0.150	55.850	55.026	0.100	54.926	54.002	5H	0	56.000	55.196	0.170	55.026	54.612	0.236	54.376
				6g	0.032	55.968	0.236	55.732	54.994	0.160	54.834	53.910	6H	0	56.000	55.238	0.212	55.026	54.676	0.300	54.376
				8g	0.032	55.968	0.375	55.593	54.994	0.250	54.744	53.820	7H	0	56.000	55.291	0.265	55.026	54.751	0.375	54.376
			2.0	4h	0	56.000	0.180	55.820	54.701	0.112	54.589	53.357	5H	0	56.000	54.891	0.190	54.701	54.135	0.300	53.835
				6g	0.038	55.962	0.280	55.682	54.663	0.180	54.483	53.251	6H	0	56.000	54.937	0.236	54.701	54.210	0.375	53.835
				8g	0.038	55.962	0.450	55.512	54.663	0.280	54.383	53.151	7H	0	56.000	55.001	0.300	54.701	54.310	0.475	53.835
			3.0	4h	0	56.000	0.236	55.764	54.051	0.132	53.919	52.071	5H	0	56.000	54.275	0.224	54.051	53.152	0.400	52.752
				6g	0.048	55.952	0.375	55.577	54.003	0.212	53.791	51.943	6H	0	56.000	54.331	0.280	54.051	53.252	0.500	52.752
				8g	0.048	55.952	0.600	55.352	54.003	0.335	53.668	51.820	7H	0	56.000	54.406	0.355	54.051	53.382	0.630	52.752
			4.0	4h	0	56.000	0.300	55.700	53.402	0.150	53.252	50.788	5H	0	56.000	53.652	0.250	53.402	52.145	0.475	51.670
				6g	0.060	55.940	0.475	55.465	53.342	0.236	53.106	50.642	6H	0	56.000	53.717	0.315	53.402	52.270	0.600	51.670
				8g	0.060	55.940	0.750	55.190	53.342	0.375	52.967	50.503	7H	0	56.000	53.802	0.400	53.402	52.430	0.750	51.670
			5.5	4h	0	56.000	0.355	55.645	52.428	0.170	52.258	48.870	5H	0	56.000	52.708	0.280	52.428	50.646	0.600	50.046
				6g	0.075	55.925	0.560	55.365	52.353	0.265	52.088	48.700	6H	0	56.000	52.783	0.355	52.428	50.796	0.750	50.046
				8g	0.075	55.925	0.900	55.025	52.353	0.425	51.928	48.540	7H	0	56.000	52.878	0.450	52.428	50.996	0.950	50.046

1	2	3	4	5	6	7	8	9	10	11	12	13	14	15	16	17	18	19	20	21	22
Nominal diameter	Pitch			External threads									Internal threads								
	Coarse	Fine	Constant	Tolerance class	Fund dev.	Major diameter			Pitch diameter			Minor diameter	Tolerance class	Fund dev.	Major diameter	Pitch diameter			Minor diameter		
						max.	tol.	min.	max.	tol.	min.	min.			min.	max.	tol.	min.	max.	tol.	min.
58			1.5	4h	0	58.000	0.150	57.850	57.026	0.100	56.926	56.002	5H	0	58.000	57.196	0.170	57.026	56.612	0.236	56.376
				6g	0.032	57.968	0.236	57.732	56.994	0.160	56.834	55.910	6H	0	58.000	57.238	0.212	57.026	56.676	0.300	56.376
				8g	0.032	57.968	0.375	57.593	56.994	0.250	56.744	55.820	7H	0	58.000	57.291	0.265	57.026	56.751	0.375	56.376
			2.0	4h	0	58.000	0.180	57.820	56.701	0.112	56.589	55.357	5H	0	58.000	56.891	0.190	56.701	56.135	0.300	55.835
				6g	0.038	57.962	0.280	57.682	56.663	0.180	56.483	55.251	6H	0	58.000	56.937	0.236	56.701	56.210	0.375	55.835
				8g	0.038	57.962	0.450	57.512	56.663	0.280	56.383	55.151	7H	0	58.000	57.001	0.300	56.701	56.310	0.475	55.835
			3.0	4h	0	58.000	0.236	57.764	56.051	0.132	55.919	54.071	5H	0	58.000	56.275	0.224	56.051	55.152	0.400	54.752
				6g	0.048	57.952	0.375	57.577	56.003	0.212	55.791	53.943	6H	0	58.000	56.331	0.280	56.051	55.252	0.500	54.752
				8g	0.048	57.952	0.600	57.352	56.003	0.335	55.668	53.820	7H	0	58.000	56.406	0.355	56.051	55.382	0.630	54.752
			4.0	4h	0	58.000	0.300	57.700	55.402	0.150	55.252	52.788	5H	0	58.000	55.652	0.250	55.402	54.145	0.475	53.670
				6g	0.060	57.940	0.475	57.465	55.342	0.236	55.106	52.642	6H	0	58.000	55.717	0.315	55.402	54.270	0.600	53.670
				8g	0.060	57.940	0.750	57.190	55.342	0.375	54.967	52.503	7H	0	58.000	55.802	0.400	55.402	54.420	0.750	53.670
60			1.5	4h	0	60.000	0.150	59.850	59.026	0.100	58.926	58.002	5H	0	60.000	59.196	0.170	59.026	58.612	0.236	58.376
				6g	0.032	59.968	0.236	59.732	58.994	0.160	58.834	57.910	6H	0	60.000	59.238	0.212	59.026	58.676	0.300	58.376
				8g	0.032	59.968	0.375	59.593	58.994	0.250	58.744	57.820	7H	0	60.000	59.291	0.265	59.026	58.751	0.375	58.376
			2.0	4h	0	60.000	0.180	59.820	58.701	0.112	58.589	57.357	5H	0	60.000	58.891	0.190	58.701	58.135	0.300	57.835
				6g	0.038	59.962	0.280	59.682	58.663	0.180	58.483	57.251	6H	0	60.000	58.937	0.236	58.701	58.210	0.375	57.835
				8g	0.038	59.962	0.450	59.512	58.663	0.280	58.383	57.151	7H	0	60.000	59.001	0.300	58.701	58.310	0.475	57.835
			3.0	4h	0	60.000	0.236	59.764	58.051	0.132	57.919	56.071	5H	0	60.000	58.275	0.224	58.051	57.152	0.400	56.752
				6g	0.048	59.952	0.375	59.577	58.003	0.212	57.791	55.943	6H	0	60.000	58.331	0.280	58.051	57.252	0.500	56.752
				8g	0.048	59.952	0.600	59.352	58.003	0.335	57.668	55.820	7H	0	60.000	58.406	0.355	58.051	57.382	0.630	56.752
			4.0	4h	0	60.000	0.300	59.700	57.402	0.150	57.252	54.788	5H	0	60.000	57.652	0.250	57.402	56.145	0.475	55.670
				6g	0.060	59.940	0.475	59.465	57.342	0.236	57.106	54.642	6H	0	60.000	57.717	0.315	57.402	56.270	0.600	55.670
				8g	0.060	59.940	0.750	59.190	57.342	0.375	56.967	54.503	7H	0	60.000	57.802	0.400	57.402	56.420	0.750	55.670
	5.5			4h	0	60.000	0.355	59.645	56.428	0.170	56.258	52.870	5H	0	60.000	56.708	0.280	56.428	54.646	0.600	54.046
				6g	0.075	59.925	0.560	59.365	56.353	0.265	56.088	52.700	6H	0	60.000	56.783	0.355	56.428	54.796	0.750	54.046
				8g	0.075	59.925	0.900	59.025	56.353	0.425	55.928	52.540	7H	0	60.000	56.878	0.450	56.428	54.996	0.950	54.046

Metric screw threads

1	2	3	4	5	6	7	8	9	10	11	12	13	14	15	16	17	18	19	20	21	22
Nominal diameter	Pitch			External threads									Internal threads								
	Coarse	Fine	Constant	Tolerance class	Fund dev.	Major diameter			Pitch diameter			Minor diameter	Tolerance class	Fund dev.	Major diameter	Pitch diameter			Minor diameter		
						max.	tol.	min.	max.	tol.	min.	min.			min.	max.	tol.	min.	max.	tol.	min.
62			1.5	4h	0	62.000	0.150	61.850	61.026	0.100	60.926	60.002	5H	0	62.000	61.196	0.170	61.026	60.612	0.236	60.376
				6g	0.032	61.968	0.236	61.732	60.994	0.160	60.834	59.910	6H	0	62.000	61.238	0.212	61.026	60.676	0.300	60.376
				8g	0.032	61.968	0.375	61.593	60.994	0.250	60.744	59.820	7H	0	62.000	61.291	0.265	61.026	60.751	0.375	60.376
			2	4h	0	62.000	0.180	61.820	60.701	0.112	60.589	59.357	5H	0	62.000	60.891	0.190	60.701	60.135	0.300	59.835
				6g	0.038	61.962	0.280	61.682	60.663	0.180	60.483	59.251	6H	0	62.000	60.937	0.236	60.701	60.210	0.375	59.835
				8g	0.038	61.962	0.450	61.512	60.663	0.280	60.383	59.151	7H	0	62.000	61.001	0.300	60.701	60.310	0.475	59.835
			3	4h	0	62.000	0.236	61.764	60.051	0.132	59.919	58.071	5H	0	62.000	60.275	0.224	60.051	59.152	0.400	58.752
				6g	0.048	61.952	0.375	61.577	60.003	0.212	59.791	57.943	6H	0	62.000	60.331	0.280	60.051	59.252	0.500	58.752
				8g	0.048	61.952	0.600	61.352	60.003	0.335	59.668	57.820	7H	0	62.000	60.406	0.355	60.051	59.382	0.630	58.752
			4	4h	0	62.000	0.300	61.700	59.402	0.150	59.252	56.788	5H	0	62.000	59.652	0.250	59.402	58.145	0.475	57.670
				6g	0.060	61.940	0.475	61.465	59.342	0.236	59.106	56.642	6H	0	62.000	59.717	0.315	59.402	58.270	0.600	57.670
				8g	0.060	61.940	0.750	61.190	59.342	0.375	58.967	56.503	7H	0	62.000	59.802	0.400	59.402	58.420	0.750	57.670
64			1.5	4h	0	64.000	0.150	63.850	63.026	0.100	62.926	62.002	5H	0	64.000	63.196	0.170	63.026	62.612	0.236	62.376
				6g	0.032	63.968	0.236	63.732	62.994	0.160	62.834	61.910	6H	0	64.000	63.238	0.212	63.026	62.676	0.300	62.376
				8g	0.032	63.968	0.375	63.593	62.994	0.250	62.744	61.820	7H	0	64.000	63.291	0.265	63.026	62.751	0.375	62.376
			2	4h	0	64.000	0.180	63.820	62.701	0.112	62.589	61.357	5H	0	64.000	62.891	0.190	62.701	62.135	0.300	61.835
				6g	0.038	63.962	0.280	63.682	62.663	0.180	62.483	61.251	6H	0	64.000	62.937	0.236	62.701	62.210	0.375	61.835
				8g	0.038	63.962	0.450	63.512	62.663	0.280	62.383	61.151	7H	0	64.000	63.001	0.300	62.701	62.310	0.475	61.835
			3	4h	0	64.000	0.236	63.764	62.051	0.132	61.919	60.071	5H	0	64.000	62.275	0.224	62.051	61.152	0.400	60.752
				6g	0.048	63.952	0.375	63.577	62.003	0.212	61.791	59.943	6H	0	64.000	62.331	0.280	62.051	61.252	0.500	60.752
				8g	0.048	63.952	0.600	63.352	62.003	0.335	61.668	59.820	7H	0	64.000	62.406	0.355	62.051	61.382	0.630	60.752
			4	4h	0	64.000	0.300	63.700	61.402	0.150	61.252	58.788	5H	0	64.000	61.652	0.250	61.402	60.145	0.475	59.670
				6g	0.060	63.940	0.475	63.465	61.342	0.236	61.106	58.642	6H	0	64.000	61.717	0.315	61.402	60.270	0.600	59.670
				8g	0.060	63.940	0.750	63.190	61.342	0.375	60.967	58.503	7H	0	64.000	61.802	0.400	61.402	60.420	0.750	59.670
	6			4h	0	64.000	0.375	63.625	60.103	0.180	59.923	56.227	5H	0	64.000	60.403	0.300	60.103	58.135	0.630	57.505
				6g	0.080	63.920	0.600	63.320	60.023	0.280	59.743	56.047	6H	0	64.000	60.478	0.375	60.103	58.305	0.800	57.505
				8g	0.080	63.920	0.950	62.970	60.023	0.450	59.573	55.877	7H	0	64.000	60.578	0.475	60.103	58.505	1.000	57.505

1	2	3	4	5	6	7	8	9	10	11	12	13	14	15	16	17	18	19	20	21	22
Nominal diameter	Pitch			External threads									Internal threads								
	Coarse	Fine	Constant	Tolerance class	Fund dev.	Major diameter			Pitch diameter			Minor diameter	Tolerance class	Fund dev.	Major diameter	Pitch diameter			Minor diameter		
						max.	tol.	min.	max.	tol.	min.	min.			min.	max.	tol.	min.	max.	tol.	min.
65			1.5	4h	0	65.000	0.150	64.850	64.026	0.100	63.926	63.002	5H	0	65.000	64.196	0.170	64.026	63.612	0.236	63.376
				6g	0.032	64.968	0.236	64.732	63.994	0.160	63.834	62.910	6H	0	65.000	64.238	0.212	64.026	63.676	0.300	63.376
				8g	0.032	64.968	0.375	64.593	63.994	0.250	63.744	62.820	7H	0	65.000	64.291	0.265	64.026	63.751	0.375	63.376
			2	4h	0	65.000	0.180	64.820	63.701	0.112	63.589	62.357	5H	0	65.000	63.891	0.190	63.701	63.135	0.300	62.835
				6g	0.038	64.962	0.280	64.682	63.663	0.180	63.483	62.251	6H	0	65.000	63.937	0.236	63.701	63.210	0.375	62.835
				8g	0.038	64.962	0.450	64.512	63.663	0.280	63.383	62.151	7H	0	65.000	64.001	0.300	63.701	63.310	0.475	62.835
			3	4h	0	65.000	0.236	64.764	63.051	0.132	62.919	61.071	5H	0	65.000	63.275	0.224	63.051	62.152	0.400	61.752
				6g	0.048	64.952	0.375	64.577	63.003	0.212	62.791	60.943	6H	0	65.000	63.331	0.280	63.051	62.252	0.500	61.752
				8g	0.048	64.952	0.600	64.352	63.003	0.335	62.668	60.820	7H	0	65.000	63.406	0.355	63.051	62.382	0.630	61.752
			4	4h	0	65.000	0.300	64.700	62.402	0.150	62.252	59.788	5H	0	65.000	62.652	0.250	62.402	61.145	0.475	60.670
				6g	0.060	64.940	0.475	64.465	62.342	0.236	62.106	59.642	6H	0	65.000	62.717	0.315	62.402	61.270	0.600	60.670
				8g	0.060	64.940	0.750	64.190	62.342	0.375	61.967	59.503	7H	0	65.000	62.802	0.400	62.402	61.420	0.750	60.670
68			1.5	4h	0	68.000	0.150	67.850	67.026	0.100	66.926	66.002	5H	0	68.000	67.196	0.170	67.026	66.612	0.236	66.376
				6g	0.032	67.968	0.236	67.732	66.994	0.160	66.834	65.910	6H	0	68.000	67.238	0.212	67.026	66.676	0.300	66.376
				8g	0.032	67.968	0.375	67.593	66.994	0.250	66.744	65.820	7H	0	68.000	67.291	0.265	67.026	66.751	0.375	66.376
			2	4h	0	68.000	0.180	67.820	66.701	0.112	66.589	65.357	5H	0	68.000	66.891	0.190	66.701	66.135	0.300	65.835
				6g	0.038	67.962	0.280	67.682	66.663	0.180	66.483	65.251	6H	0	68.000	66.937	0.236	66.701	66.210	0.375	65.835
				8g	0.038	67.962	0.450	67.512	66.663	0.280	66.383	65.151	7H	0	68.000	67.001	0.300	66.701	66.310	0.475	65.835
			3	4h	0	68.000	0.236	67.764	66.051	0.132	65.919	64.071	5H	0	68.000	66.275	0.224	66.051	65.152	0.400	64.752
				6g	0.048	67.952	0.375	67.577	66.003	0.212	65.791	63.943	6H	0	68.000	66.331	0.280	66.051	65.252	0.500	64.752
				8g	0.048	67.952	0.600	67.352	66.003	0.335	65.668	63.820	7H	0	68.000	66.406	0.355	66.051	65.382	0.630	64.752
			4	4h	0	68.000	0.300	67.700	65.402	0.150	65.252	62.788	5H	0	68.000	65.652	0.250	65.402	64.145	0.475	63.670
				6g	0.060	67.940	0.475	67.465	65.342	0.236	65.106	62.642	6H	0	68.000	65.717	0.315	65.402	64.270	0.600	63.670
				8g	0.060	67.940	0.750	67.190	65.342	0.375	64.967	62.503	7H	0	68.000	65.802	0.400	65.402	64.420	0.750	63.670
		6		4h	0	68.000	0.375	67.625	64.103	0.180	63.923	60.227	5H	0	68.000	64.403	0.300	64.103	62.135	0.630	61.505
				6g	0.080	67.920	0.600	67.320	64.023	0.280	63.743	60.047	6H	0	68.000	64.478	0.375	64.103	62.305	0.800	61.505
				8g	0.080	67.920	0.950	66.970	64.023	0.450	63.573	59.877	7H	0	68.000	64.578	0.475	64.103	62.505	1.000	61.505

… Metric screw threads

1	2	3	4	5	6	7	8	9	10	11	12	13	14	15	16	17	18	19	20	21	22
Nominal diameter	Pitch			External threads									Internal threads								
	Coarse	Fine	Constant	Tolerance class	Fund dev.	Major diameter			Pitch diameter			Minor diameter	Tolerance class	Fund dev.	Major diameter	Pitch diameter			Minor diameter		
						max.	tol.	min.	max.	tol.	min.	min.			min.	max.	tol.	min.	max.	tol.	min.
70			1.5	4h	0	70.000	0.150	69.850	69.026	0.100	68.926	68.002	5H	0	70.000	69.196	0.170	69.026	68.612	0.236	68.376
				6g	0.032	69.968	0.236	69.732	68.994	0.160	68.834	67.910	6H	0	70.000	69.238	0.212	69.026	68.676	0.300	68.376
				8g	0.032	69.968	0.375	69.593	68.994	0.250	68.744	67.820	7H	0	70.000	69.291	0.265	69.026	68.751	0.375	68.376
			2	4h	0	70.000	0.180	69.820	68.701	0.112	68.589	67.357	5H	0	70.000	68.891	0.190	68.701	68.135	0.300	67.835
				6g	0.038	69.962	0.280	69.682	68.663	0.180	68.483	67.251	6H	0	70.000	68.937	0.236	68.701	68.210	0.375	67.835
				8g	0.038	69.962	0.450	69.512	68.663	0.280	68.383	67.151	7H	0	70.000	69.001	0.300	68.701	68.310	0.475	67.835
			3	4h	0	70.000	0.236	69.764	68.051	0.132	67.919	66.071	5H	0	70.000	68.275	0.224	68.051	67.152	0.400	66.752
				6g	0.048	69.952	0.375	69.577	68.003	0.212	67.791	65.943	6H	0	70.000	68.331	0.280	68.051	67.252	0.500	66.752
				8g	0.048	69.952	0.600	69.352	68.003	0.335	67.668	65.820	7H	0	70.000	68.406	0.355	68.051	67.382	0.630	66.752
			4	4h	0	70.000	0.300	69.700	67.402	0.150	67.252	64.788	5H	0	70.000	67.652	0.250	67.402	66.145	0.475	65.670
				6g	0.060	69.940	0.475	69.465	67.342	0.236	67.106	64.642	6H	0	70.000	67.717	0.315	67.402	66.270	0.600	65.670
				8g	0.060	69.940	0.750	69.190	67.342	0.375	66.967	64.503	7H	0	70.000	67.802	0.400	67.402	66.420	0.750	65.670
			6	4h	0	70.000	0.375	69.625	66.103	0.180	65.923	62.227	5H	0	70.000	66.403	0.300	66.103	64.135	0.630	63.505
				6g	0.080	69.920	0.600	69.320	66.023	0.280	65.743	62.047	6H	0	70.000	66.478	0.375	66.103	64.305	0.800	63.505
				8g	0.080	69.920	0.950	68.970	66.023	0.450	65.573	61.877	7H	0	70.000	66.578	0.475	66.103	64.505	1.000	63.505
72			1.5	4h	0	72.000	0.150	71.850	71.026	0.100	70.926	70.002	5H	0	72.000	71.196	0.170	71.026	70.612	0.236	70.376
				6g	0.032	71.968	0.236	71.732	70.994	0.160	70.834	69.910	6H	0	72.000	71.238	0.212	71.026	70.676	0.300	70.376
				8g	0.032	71.968	0.375	71.593	70.994	0.250	70.744	69.820	7H	0	72.000	71.291	0.265	71.026	70.751	0.375	70.376
			2	4h	0	72.000	0.180	71.820	70.701	0.112	70.589	69.357	5H	0	72.000	70.891	0.190	70.701	70.135	0.300	69.835
				6g	0.038	71.962	0.280	71.682	70.663	0.180	70.483	69.251	6H	0	72.000	70.937	0.236	70.701	70.210	0.375	69.835
				8g	0.038	71.962	0.450	71.512	70.663	0.280	70.383	69.151	7H	0	72.000	71.001	0.300	70.701	70.310	0.475	69.835
			3	4h	0	72.000	0.236	71.764	70.051	0.132	69.919	68.071	5H	0	72.000	70.275	0.224	70.051	69.152	0.400	68.752
				6g	0.048	71.952	0.375	71.577	70.003	0.212	69.791	67.943	6H	0	72.000	70.331	0.280	70.051	69.252	0.500	68.752
				8g	0.048	71.952	0.600	71.352	70.003	0.335	69.668	67.820	7H	0	72.000	70.406	0.355	70.051	69.382	0.630	68.752
			4	4h	0	72.000	0.300	71.700	69.402	0.150	69.252	66.788	5H	0	72.000	69.652	0.250	69.402	68.145	0.475	67.670
				6g	0.060	71.940	0.475	71.465	69.342	0.236	69.106	66.642	6H	0	72.000	69.717	0.315	69.402	68.270	0.600	67.670
				8g	0.060	71.940	0.750	71.190	69.342	0.375	68.967	66.503	7H	0	72.000	69.802	0.400	69.402	68.420	0.750	67.670
			6	4h	0	72.000	0.375	71.625	68.103	0.180	67.923	64.227	5H	0	72.000	68.403	0.300	68.103	66.135	0.630	65.505
				6g	0.080	71.920	0.600	71.320	68.023	0.280	67.743	64.047	6H	0	72.000	68.478	0.375	68.103	66.305	0.800	65.505
				8g	0.080	71.920	0.950	70.970	68.023	0.450	67.573	63.877	7H	0	72.000	68.578	0.475	68.103	66.505	1.000	65.505

1	2	3	4	5	6	7	8	9	10	11	12	13	14	15	16	17	18	19	20	21	22
Nominal diameter	Pitch			External threads									Internal threads								
	Coarse	Fine	Constant	Tolerance class	Fund dev.	Major diameter			Pitch diameter			Minor diameter	Tolerance class	Fund dev.	Major diameter	Pitch diameter			Minor diameter		
						max.	tol.	min.	max.	tol.	min.	min.			min.	max.	tol.	min.	max.	tol.	min.
75			1.5	4h	0	75.000	0.150	74.850	74.026	0.100	73.926	73.002	5H	0	75.000	74.196	0.170	74.026	73.612	0.236	73.376
				6g	0.032	74.968	0.236	74.732	73.994	0.160	73.834	72.910	6H	0	75.000	74.238	0.212	74.026	73.676	0.300	73.376
				8g	0.032	74.968	0.375	74.593	73.994	0.250	73.744	72.820	7H	0	75.000	74.291	0.265	74.026	73.751	0.375	73.376
			2	4h	0	75.000	0.180	74.820	73.701	0.112	73.589	72.357	5H	0	75.000	73.891	0.190	73.701	73.135	0.300	72.835
				6g	0.038	74.962	0.280	74.682	73.663	0.180	73.483	72.251	6H	0	75.000	73.937	0.236	73.701	73.210	0.375	72.835
				8g	0.038	74.962	0.450	74.512	73.663	0.280	73.383	72.151	7H	0	75.000	74.001	0.300	73.701	73.310	0.475	72.835
			3	4h	0	75.000	0.236	74.764	73.051	0.132	72.919	71.071	5H	0	75.000	73.275	0.224	73.051	72.152	0.400	71.752
				6g	0.048	74.952	0.375	74.577	73.003	0.212	72.791	70.943	6H	0	75.000	73.331	0.280	73.051	72.252	0.500	71.752
				8g	0.048	74.952	0.600	74.352	73.003	0.335	72.668	70.820	7H	0	75.000	73.406	0.355	73.051	72.382	0.630	71.752
			4	4h	0	75.000	0.300	74.700	72.402	0.150	72.252	69.788	5H	0	75.000	72.652	0.250	72.402	71.145	0.475	70.670
				6g	0.060	74.940	0.475	74.465	72.342	0.236	72.106	69.642	6H	0	75.000	72.717	0.315	72.402	71.270	0.600	70.670
				8g	0.060	74.940	0.750	74.190	72.342	0.375	71.967	69.503	7H	0	75.000	72.802	0.400	72.402	71.420	0.750	70.670
76			1.5	4h	0	76.000	0.150	75.850	75.026	0.100	74.926	74.002	5H	0	76.000	75.196	0.170	75.026	74.612	0.236	74.376
				6g	0.032	75.968	0.236	75.732	74.994	0.160	74.834	73.910	6H	0	76.000	75.238	0.212	75.026	74.676	0.300	74.376
				8g	0.032	75.968	0.375	75.593	74.994	0.250	74.744	73.820	7H	0	76.000	75.291	0.265	75.026	74.751	0.375	74.376
			2	4h	0	76.000	0.180	75.820	74.701	0.112	74.589	73.357	5H	0	76.000	74.891	0.190	74.701	74.135	0.300	73.835
				6g	0.038	75.962	0.280	75.682	74.663	0.180	74.483	73.251	6H	0	76.000	74.937	0.236	74.701	74.210	0.375	73.835
				8g	0.038	75.962	0.450	75.512	74.663	0.280	74.383	73.151	7H	0	76.000	75.001	0.300	74.701	74.310	0.475	73.835
			3	4h	0	76.000	0.236	75.764	74.051	0.132	73.919	72.071	5H	0	76.000	74.275	0.224	74.051	73.152	0.400	72.752
				6g	0.048	75.952	0.375	75.577	74.003	0.212	73.791	71.943	6H	0	76.000	74.331	0.280	74.051	73.252	0.500	72.752
				8g	0.048	75.952	0.600	75.352	74.003	0.335	73.668	71.820	7H	0	76.000	74.406	0.355	74.051	73.382	0.630	72.752
			4	4h	0	76.000	0.300	75.700	73.402	0.150	73.252	70.788	5H	0	76.000	73.652	0.250	73.402	72.145	0.475	71.670
				6g	0.060	75.940	0.475	75.465	73.342	0.236	73.106	70.642	6H	0	76.000	73.717	0.315	73.402	72.270	0.600	71.670
				8g	0.060	75.940	0.750	75.190	73.342	0.375	72.967	70.503	7H	0	76.000	73.802	0.400	73.402	72.420	0.750	71.670
			6	4h	0	76.000	0.375	75.625	72.103	0.180	71.923	68.227	5H	0	76.000	72.403	0.300	72.103	70.135	0.630	69.505
				6g	0.080	75.920	0.600	75.320	72.023	0.280	71.743	68.047	6H	0	76.000	72.478	0.375	72.103	70.305	0.800	69.505
				8g	0.080	75.920	0.950	74.970	72.023	0.450	71.573	67.877	7H	0	76.000	72.578	0.475	72.103	70.505	1.000	69.505
78			2	4h	0	78.000	0.180	77.820	76.701	0.112	76.589	75.357	5H	0	78.000	76.891	0.190	76.701	76.135	0.300	75.835
				6g	0.038	77.962	0.280	77.682	76.663	0.180	76.483	75.251	6H	0	78.000	76.937	0.236	76.701	76.210	0.375	75.835
				8g	0.038	77.962	0.450	77.512	76.663	0.280	76.383	75.151	7H	0	78.000	77.001	0.300	76.701	76.310	0.475	75.835

Metric screw threads

1	2	3	4	5	6	7	8	9	10	11	12	13	14	15	16	17	18	19	20	21	22
Nominal diameter	Pitch			External threads									Internal threads								
	Coarse	Fine	Constant	Tolerance class	Fund dev.	Major diameter			Pitch diameter			Minor diameter	Tolerance class	Fund dev.	Major diameter	Pitch diameter			Minor diameter		
						max.	tol.	min.	max.	tol.	min.	min.			min.	max.	tol.	min.	max.	tol.	min.
80			1.5	4h	0	80.000	0.150	79.850	79.026	0.100	78.926	78.002	5H	0	80.000	79.196	0.170	79.026	78.612	0.236	78.376
				6g	0.032	79.968	0.236	79.732	78.994	0.160	78.834	77.910	6H	0	80.000	79.238	0.212	79.026	78.676	0.300	78.376
				8g	0.032	79.968	0.375	79.593	78.994	0.250	78.774	77.820	7H	0	80.000	79.291	0.265	79.026	78.751	0.375	78.376
			2	4h	0	80.000	0.180	79.820	78.701	0.112	78.589	77.357	5H	0	80.000	78.891	0.190	78.701	78.135	0.300	77.835
				6g	0.038	79.962	0.280	79.682	78.663	0.180	78.483	77.251	6H	0	80.000	78.937	0.236	78.701	78.210	0.375	77.835
				8g	0.038	79.962	0.450	79.512	78.663	0.280	78.383	77.151	7H	0	80.000	79.001	0.300	78.701	78.310	0.475	77.835
			3	4h	0	80.000	0.236	79.764	78.051	0.132	77.919	76.071	5H	0	80.000	78.275	0.224	78.051	77.152	0.400	76.752
				6g	0.048	79.952	0.375	79.577	78.003	0.212	77.791	75.943	6H	0	80.000	78.331	0.280	78.051	77.252	0.500	76.752
				8g	0.048	79.952	0.600	79.352	78.003	0.335	77.668	75.820	7H	0	80.000	78.406	0.355	78.051	77.382	0.630	76.752
			4	4h	0	80.000	0.300	79.700	77.402	0.150	77.252	74.788	5H	0	80.000	77.652	0.250	77.402	76.145	0.475	75.670
				6g	0.060	79.940	0.475	79.465	77.342	0.236	77.106	74.642	6H	0	80.000	77.717	0.315	77.402	76.270	0.600	75.670
				8g	0.060	79.940	0.750	79.190	77.342	0.375	76.967	74.503	7H	0	80.000	77.802	0.400	77.402	76.420	0.750	75.670
			6	4h	0	80.000	0.375	79.625	76.103	0.180	75.923	72.227	5H	0	80.000	76.403	0.300	76.103	74.135	0.630	73.505
				6g	0.080	79.920	0.600	79.320	76.023	0.280	75.743	72.047	6H	0	80.000	76.478	0.375	76.103	74.305	0.800	73.505
				8g	0.080	79.920	0.950	78.970	76.023	0.450	75.573	71.877	7H	0	80.000	76.578	0.475	76.103	74.505	1.000	73.505
82			2	4h	0	82.000	0.180	81.820	80.701	0.112	80.589	79.357	5H	0	82.000	80.891	0.190	80.701	80.135	0.300	79.835
				6g	0.038	81.962	0.280	81.682	80.663	0.180	80.483	79.251	6H	0	82.000	80.937	0.236	80.701	80.210	0.375	79.835
				8g	0.038	81.962	0.450	81.512	80.663	0.280	80.383	79.151	7H	0	82.000	81.001	0.300	80.701	80.310	0.475	79.835
85			2	4h	0	85.000	0.180	84.820	83.701	0.112	83.589	82.357	5H	0	85.000	83.891	0.190	83.701	83.135	0.300	82.835
				6g	0.038	84.962	0.280	84.682	83.663	0.180	83.483	82.251	6H	0	85.000	83.937	0.236	83.701	83.210	0.375	82.835
				8g	0.038	84.962	0.450	84.512	83.663	0.280	83.383	82.151	7H	0	85.000	84.001	0.300	83.701	83.310	0.475	82.835
			3	4h	0	85.000	0.236	84.764	83.051	0.132	82.919	81.071	5H	0	85.000	83.275	0.224	83.051	82.152	0.400	81.752
				6g	0.048	84.952	0.375	84.577	83.003	0.212	82.791	80.943	6H	0	85.000	83.331	0.280	83.051	82.252	0.500	81.752
				8g	0.048	84.952	0.600	84.352	83.003	0.335	82.668	80.820	7H	0	85.000	83.406	0.355	83.051	82.382	0.630	81.752
			4	4h	0	85.000	0.300	84.700	82.402	0.150	82.252	79.788	5H	0	85.000	82.652	0.250	82.402	81.145	0.475	80.670
				6g	0.060	84.940	0.475	84.465	82.342	0.236	82.106	79.642	6H	0	85.000	82.717	0.315	82.402	81.270	0.600	80.670
				8g	0.060	84.940	0.750	84.190	82.342	0.375	81.967	79.503	7H	0	85.000	82.802	0.400	82.402	81.420	0.750	80.670
			6	4h	0	85.000	0.375	84.625	81.103	0.180	80.923	77.227	5H	0	85.000	81.403	0.300	81.103	79.135	0.630	78.505
				6g	0.080	84.920	0.600	84.320	81.023	0.280	80.743	77.047	6H	0	85.000	81.478	0.375	81.103	79.305	0.800	78.505
				8g	0.080	84.920	0.950	84.970	81.023	0.450	81.573	76.877	7H	0	85.000	81.578	0.475	81.103	79.505	1.000	78.505

1	2	3	4	5	6	7	8	9	10	11	12	13	14	15	16	17	18	19	20	21	22
Nominal diameter	Pitch			External threads									Internal threads								
	Coarse	Fine	Constant	Tolerance class	Fund dev.	Major diameter			Pitch diameter			Minor diameter	Tolerance class	Fund dev.	Major diameter	Pitch diameter			Minor diameter		
						max.	tol.	min.	max.	tol.	min.	min.			min.	max.	tol.	min.	max.	tol.	min.
90			2	4h	0	90.000	0.180	89.820	88.701	0.112	88.589	87.357	5H	0	90.000	88.891	0.190	88.701	88.135	0.300	87.835
				6g	0.038	89.962	0.280	89.682	88.663	0.180	88.483	87.251	6H	0	90.000	88.937	0.236	88.701	88.210	0.375	87.835
				8g	0.038	89.962	0.450	89.512	88.663	0.280	88.383	87.151	7H	0	90.000	89.001	0.300	88.701	88.310	0.475	87.835
			3	4h	0	90.000	0.236	89.764	88.051	0.132	87.919	86.071	5H	0	90.000	88.275	0.224	88.051	87.152	0.400	86.752
				6g	0.048	89.952	0.375	89.577	88.003	0.212	87.791	85.943	6H	0	90.000	88.331	0.280	88.051	87.252	0.500	86.752
				8g	0.048	89.952	0.600	89.352	88.003	0.335	87.668	85.820	7H	0	90.000	88.406	0.355	88.051	87.382	0.630	86.752
			4	4h	0	90.000	0.300	89.700	87.402	0.150	87.252	84.788	5H	0	90.000	87.652	0.250	87.402	86.145	0.475	85.670
				6g	0.060	89.940	0.475	89.465	87.342	0.236	87.106	84.642	6H	0	90.000	87.717	0.315	87.402	86.270	0.600	85.670
				8g	0.060	89.940	0.750	89.190	87.342	0.375	86.967	84.503	7H	0	90.000	87.802	0.400	87.402	86.420	0.750	85.670
			6	4h	0	90.000	0.375	89.625	86.103	0.180	85.923	82.227	5H	0	90.000	86.403	0.300	86.103	84.135	0.630	83.505
				6g	0.080	89.920	0.600	89.320	86.023	0.280	85.743	82.047	6H	0	90.000	86.478	0.375	86.103	84.305	0.800	83.505
				8g	0.080	89.920	0.950	88.970	86.023	0.450	85.573	81.877	7H	0	90.000	86.578	0.475	86.103	84.505	1.000	83.505
95			2	4h	0	95.000	0.180	94.820	93.701	0.118	93.583	92.351	5H	0	95.000	93.901	0.200	93.701	93.135	0.300	92.835
				6g	0.038	94.962	0.280	94.682	93.663	0.190	93.473	92.241	6H	0	95.000	93.951	0.250	93.701	93.210	0.375	92.835
				8g	0.038	94.962	0.450	94.512	93.663	0.300	93.363	92.131	7H	0	95.000	94.016	0.315	93.701	93.310	0.475	92.835
			3	4h	0	95.000	0.236	94.764	93.051	0.140	92.911	91.063	5H	0	95.000	93.287	0.236	93.051	92.152	0.400	91.752
				6g	0.048	94.952	0.375	94.577	93.003	0.224	92.779	90.931	6H	0	95.000	93.351	0.300	93.051	92.252	0.500	91.752
				8g	0.048	94.952	0.600	94.352	93.003	0.355	92.648	90.800	7H	0	95.000	93.426	0.375	93.051	92.382	0.630	91.752
			4	4h	0	95.000	0.300	94.700	92.402	0.160	92.242	89.778	5H	0	95.000	92.667	0.265	92.402	91.145	0.475	90.670
				6g	0.060	94.940	0.475	94.465	92.342	0.250	92.092	89.628	6H	0	95.000	92.737	0.335	92.402	91.270	0.600	90.670
				8g	0.060	94.940	0.750	94.190	92.342	0.400	91.942	89.478	7H	0	95.000	92.827	0.425	92.402	91.420	0.750	90.670
			6	4h	0	95.000	0.375	94.625	91.103	0.190	90.913	87.217	5H	0	95.000	91.418	0.315	91.103	89.135	0.630	88.505
				6g	0.080	94.920	0.600	94.320	91.023	0.300	90.723	87.027	6H	0	95.000	91.503	0.400	91.103	89.305	0.800	88.505
				8g	0.080	94.920	0.950	93.970	91.023	0.475	90.548	86.852	7H	0	95.000	91.603	0.500	91.103	89.505	1.000	88.505

Metric screw threads

1	2	3	4	5	6	7	8	9	10	11	12	13	14	15	16	17	18	19	20	21	22
Nominal diameter	Pitch			External threads									Internal threads								
	Coarse	Fine	Constant	Tolerance class	Fund dev.	Major diameter			Pitch diameter			Minor diameter	Tolerance class	Fund dev.	Major diameter	Pitch diameter			Minor diameter		
						max.	tol.	min.	max.	tol.	min.	min.			min.	max.	tol.	min.	max.	tol.	min.
100			2	4h	0	100.000	0.180	99.820	98.701	0.118	98.583	97.351	5H	0	100.000	98.901	0.200	98.701	98.135	0.300	97.835
				6g	0.038	99.962	0.280	99.682	98.663	0.190	98.473	97.241	6H	0	100.000	98.951	0.250	98.701	98.210	0.375	97.835
				8g	0.038	99.962	0.450	99.512	98.663	0.300	98.363	97.131	7H	0	100.000	99.016	0.315	98.701	98.310	0.475	97.835
			3	4h	0	100.000	0.236	99.764	98.051	0.140	97.911	96.063	5H	0	100.000	98.287	0.236	98.051	97.152	0.400	96.752
				6g	0.048	99.952	0.375	99.577	98.003	0.224	97.779	95.931	6H	0	100.000	98.351	0.300	98.051	97.252	0.500	96.752
				8g	0.048	99.952	0.600	99.352	98.003	0.355	97.648	95.800	7H	0	100.000	98.426	0.375	98.051	97.382	0.630	96.752
			4	4h	0	100.000	0.300	99.700	97.402	0.160	97.242	94.778	5H	0	100.000	97.667	0.265	97.402	96.145	0.475	95.670
				6g	0.060	99.940	0.475	99.465	97.342	0.250	97.092	94.628	6H	0	100.000	97.737	0.335	97.402	96.270	0.600	95.670
				8g	0.060	99.940	0.750	99.190	97.342	0.400	96.942	94.478	7H	0	100.000	97.827	0.425	97.402	96.420	0.750	95.670
			6	4h	0	100.000	0.375	99.625	96.103	0.190	95.913	92.217	5H	0	100.000	96.418	0.315	96.103	94.135	0.630	93.505
				6g	0.080	99.920	0.600	99.320	96.023	0.300	95.723	92.027	6H	0	100.000	96.503	0.400	96.103	94.305	0.800	93.505
				8g	0.080	99.920	0.950	98.970	96.023	0.475	95.548	91.852	7H	0	100.000	96.603	0.500	96.103	94.505	1.000	93.505
105			2	4h	0	105.000	0.180	104.820	103.701	0.118	103.583	102.351	5H	0	105.000	103.901	0.200	103.701	103.135	0.300	102.835
				6g	0.038	104.962	0.280	104.682	103.663	0.190	103.473	102.241	6H	0	105.000	103.951	0.250	103.701	103.210	0.375	102.835
				8g	0.038	104.962	0.450	104.512	103.663	0.300	103.363	102.131	7H	0	105.000	104.016	0.315	103.701	103.310	0.475	102.835
			3	4h	0	105.000	0.236	104.764	103.051	0.140	102.911	101.063	5H	0	105.000	103.287	0.236	103.051	102.152	0.400	101.752
				6g	0.048	104.952	0.375	104.577	103.003	0.224	102.779	100.931	6H	0	105.000	103.351	0.300	103.051	102.252	0.500	101.752
				8g	0.048	104.952	0.600	104.352	103.003	0.355	102.648	100.800	7H	0	105.000	103.426	0.375	103.051	102.382	0.630	101.752
			4	4h	0	105.000	0.300	104.700	102.402	0.160	102.242	99.778	5H	0	105.000	102.667	0.265	102.402	101.145	0.475	100.670
				6g	0.060	104.940	0.475	104.465	102.342	0.250	102.092	99.628	6H	0	105.000	102.737	0.335	102.402	101.270	0.600	100.670
				8g	0.060	104.940	0.750	104.190	102.342	0.400	101.942	99.478	7H	0	105.000	102.827	0.425	102.402	101.420	0.750	100.670
			6	4h	0	105.000	0.375	104.625	101.103	0.190	100.913	97.217	5H	0	105.000	101.418	0.315	101.103	99.135	0.630	98.505
				6g	0.080	104.920	0.600	104.320	101.023	0.300	100.723	97.027	6H	0	105.000	101.503	0.400	101.103	99.305	0.800	98.505
				8g	0.080	104.920	0.950	103.970	101.023	0.475	100.548	96.852	7H	0	105.000	101.603	0.500	101.103	99.505	1.000	98.505

1	2	3	4	5	6	7	8	9	10	11	12	13	14	15	16	17	18	19	20	21	22
Nominal diameter	Pitch			External threads									Internal threads								
	Coarse	Fine	Constant	Tolerance class	Fund dev.	Major diameter			Pitch diameter			Minor diameter	Tolerance class	Fund dev.	Major diameter	Pitch diameter			Minor diameter		
						max.	tol.	min.	max.	tol.	min.	min.			min.	max.	tol.	min.	max.	tol.	min.
110			2	4h	0	110.000	0.180	109.820	108.701	0.118	108.583	107.351	5H	0	110.000	108.901	0.200	108.701	108.135	0.300	107.835
				6g	0.038	109.962	0.280	109.682	108.663	0.190	108.473	107.241	6H	0	110.000	108.951	0.250	108.701	108.210	0.375	107.835
				8g	0.038	109.962	0.450	109.512	108.663	0.300	108.363	107.131	7H	0	110.000	109.016	0.315	108.701	108.310	0.475	107.835
			3	4h	0	110.000	0.236	109.764	108.051	0.140	107.911	106.063	5H	0	110.000	108.287	0.236	108.051	107.152	0.400	106.752
				6g	0.048	109.952	0.375	109.577	108.003	0.224	107.779	105.931	6H	0	110.000	108.351	0.300	108.051	107.252	0.500	106.752
				8g	0.048	109.952	0.600	109.352	108.003	0.355	107.648	105.800	7H	0	110.000	108.426	0.375	108.051	107.382	0.630	106.752
			4	4h	0	110.000	0.300	109.700	107.402	0.160	107.242	104.778	5H	0	110.000	107.667	0.265	107.402	106.145	0.475	105.670
				6g	0.060	109.940	0.475	109.465	107.342	0.250	107.092	104.628	6H	0	110.000	107.737	0.335	107.402	106.270	0.600	105.670
				8g	0.060	109.940	0.750	109.190	107.342	0.400	106.942	104.478	7H	0	110.000	107.827	0.425	107.402	106.420	0.750	105.670
			6	4h	0	110.000	0.375	109.625	106.103	0.190	105.913	102.217	5H	0	110.000	106.418	0.315	106.103	104.135	0.630	103.505
				6g	0.080	109.920	0.600	109.320	106.023	0.300	105.723	102.027	6H	0	110.000	106.503	0.400	106.103	104.305	0.800	103.505
				8g	0.080	109.920	0.950	108.970	106.023	0.475	105.548	101.852	7H	0	110.000	106.603	0.500	106.103	104.505	1.000	103.505
115			2	4h	0	115.000	0.180	114.820	113.701	0.118	113.583	112.351	5H	0	115.000	113.901	0.200	113.701	113.135	0.300	112.835
				6g	0.038	114.962	0.280	114.682	113.663	0.190	113.473	112.241	6H	0	115.000	113.951	0.250	113.701	113.210	0.375	112.835
				8g	0.038	114.962	0.450	114.512	113.663	0.300	113.363	112.131	7H	0	115.000	114.016	0.315	113.701	113.310	0.475	112.835
			3	4h	0	115.000	0.236	114.764	113.051	0.140	112.911	111.063	5H	0	115.000	113.287	0.236	113.051	112.152	0.400	111.752
				6g	0.048	114.952	0.375	114.577	113.003	0.224	112.779	110.931	6H	0	115.000	113.351	0.300	113.051	112.252	0.500	111.752
				8g	0.048	114.952	0.600	114.352	113.003	0.355	112.648	110.800	7H	0	115.000	113.426	0.375	113.051	112.382	0.630	111.752
			4	4h	0	115.000	0.300	114.700	112.402	0.160	112.242	109.778	5H	0	115.000	112.667	0.265	112.402	111.145	0.475	110.670
				6g	0.060	114.940	0.475	114.465	112.342	0.250	112.092	109.628	6H	0	115.000	112.737	0.335	112.402	111.270	0.600	110.670
				8g	0.060	114.940	0.750	114.190	112.342	0.400	111.942	109.478	7H	0	115.000	112.827	0.425	112.402	111.420	0.750	110.670
			6	4h	0	115.000	0.375	114.625	111.103	0.190	110.913	107.217	5H	0	115.000	111.418	0.315	111.103	109.135	0.630	108.505
				6g	0.080	114.920	0.600	114.320	111.023	0.300	110.723	107.027	6H	0	115.000	111.503	0.400	111.103	109.305	0.800	108.505
				8g	0.080	114.920	0.950	113.970	111.023	0.475	110.548	106.852	7H	0	115.000	111.603	0.500	111.103	109.505	1.000	108.505

Metric screw threads

1	2	3	4	5	6	7	8	9	10	11	12	13	14	15	16	17	18	19	20	21	22
Nominal diameter	Pitch			External threads									Internal threads								
	Coarse	Fine	Constant	Tolerance class	Fund dev.	Major diameter			Pitch diameter			Minor diameter	Tolerance class	Fund dev.	Major diameter	Pitch diameter			Minor diameter		
						max.	tol.	min.	max.	tol.	min.	min.			min.	max.	tol.	min.	max.	tol.	min.
120			2	4h	0	120.000	0.180	119.820	118.701	0.118	118.583	117.351	5H	0	120.000	118.901	0.200	118.701	118.135	0.300	117.835
				6g	0.038	119.962	0.280	119.682	118.663	0.190	118.473	117.241	6H	0	120.000	118.951	0.250	118.701	118.210	0.375	117.835
				8g	0.038	119.962	0.450	119.512	118.663	0.300	118.363	117.131	7H	0	120.000	119.016	0.315	118.701	118.310	0.475	117.835
			3	4h	0	120.000	0.236	119.764	118.051	0.140	117.911	116.063	5H	0	120.000	118.287	0.236	118.051	117.152	0.400	116.752
				6g	0.048	119.952	0.375	119.577	118.003	0.224	117.779	115.931	6H	0	120.000	118.351	0.300	118.051	117.252	0.500	116.752
				8g	0.048	119.952	0.600	119.352	118.003	0.355	117.648	115.800	7H	0	120.000	118.426	0.375	118.051	117.382	0.630	116.752
			4	4h	0	120.000	0.300	119.700	117.402	0.160	117.242	114.778	5H	0	120.000	117.667	0.265	117.402	116.145	0.475	115.670
				6g	0.060	119.940	0.475	119.465	117.342	0.250	117.092	114.628	6H	0	120.000	117.737	0.335	117.402	116.270	0.600	115.670
				8g	0.060	119.940	0.750	119.190	117.342	0.400	116.942	114.478	7H	0	120.000	117.827	0.425	117.402	116.420	0.750	115.670
			6	4h	0	120.000	0.375	119.625	116.103	0.190	115.913	112.217	5H	0	120.000	116.418	0.315	116.103	114.135	0.630	113.505
				6g	0.080	119.920	0.600	119.320	116.023	0.300	115.723	112.027	6H	0	120.000	116.503	0.400	116.103	114.305	0.800	113.505
				8g	0.080	119.920	0.950	118.970	116.023	0.475	115.548	111.852	7H	0	120.000	116.603	0.500	116.103	114.505	1.000	113.505
125			2	4h	0	125.000	0.180	124.820	123.701	0.118	123.583	122.351	5H	0	125.000	123.901	0.200	123.701	123.135	0.300	122.835
				6g	0.038	124.962	0.280	124.682	123.663	0.190	123.473	122.241	6H	0	125.000	123.951	0.250	123.701	123.210	0.375	122.835
				8g	0.038	124.962	0.450	124.512	123.663	0.300	123.363	122.131	7H	0	125.000	124.016	0.315	123.701	123.310	0.475	122.835
			3	4h	0	125.000	0.236	124.764	123.051	0.140	122.911	121.063	5H	0	125.000	123.287	0.236	123.051	122.152	0.400	121.752
				6g	0.048	124.952	0.375	124.577	123.003	0.224	122.779	120.931	6H	0	125.000	123.351	0.300	123.051	122.252	0.500	121.752
				8g	0.048	124.952	0.600	124.352	123.003	0.355	122.648	120.800	7H	0	125.000	123.426	0.375	123.051	122.382	0.630	121.752
			4	4h	0	125.000	0.300	124.700	122.402	0.160	122.242	119.778	5H	0	125.000	122.667	0.265	122.402	121.145	0.475	120.670
				6g	0.060	124.940	0.475	124.465	122.342	0.250	122.092	119.628	6H	0	125.000	122.737	0.335	122.402	121.270	0.600	120.670
				8g	0.060	124.940	0.750	124.190	122.342	0.400	121.942	119.478	7H	0	125.000	122.827	0.425	122.402	121.420	0.750	120.670
			6	4h	0	125.000	0.375	124.625	121.103	0.190	120.913	117.217	5H	0	125.000	121.418	0.315	121.103	119.135	0.630	118.505
				6g	0.080	124.920	0.600	124.320	121.023	0.300	120.723	117.027	6H	0	125.000	121.503	0.400	121.103	119.305	0.800	118.505
				8g	0.080	124.920	0.950	123.970	121.023	0.475	120.548	116.852	7H	0	125.000	121.603	0.500	121.103	119.505	1.000	118.505

6.3 Relevant standards

BS 3643-1, *ISO metric screw threads — Part 1: Principles and basic data*
BS 3643-2, *ISO metric screw threads — Part 2: Specification for selected limits of size*

Chapter 7

Illustrated index to BS 8888

Normative references

Table A.1, on the following pages, is extracted (and adapted slightly) from BS 8888:2008. It lists standards containing requirements which need to be met in order to claim compliance with BS 8888. Where available, the table also gives a typical example of an illustration from each standard.

Abbreviations used in the table

GPP: General principles of presentation
GPS: Geometrical product specifications
GT: Geometrical tolerancing
HCTI: Handling of computer-based technical information
SQUS: Specification for quantities, units and symbols
STTP: Screw threads and threaded parts
TD: Technical drawings
TPD: Technical product documentation

Table A.1 Normative references

BS 8888 (sub)clause	Standard referenced	Title of the standard	Typical illustration
4.1, 4.2.4	BS EN ISO 1	Standard reference temperature for geometrical product specification and verification	Not available
12.1.2, 15.3.2	BS ISO 31-0	Quantities and units	Not available
12.1.2	BS ISO 31-1	SQUS – Part 1: Space and time	Not available
12.1.2	BS ISO 31-2	SQUS – Part 2: Periodic and related phenomena	Not available
12.1.2	BS ISO 31-3	SQUS – Part 3: Mechanics	Not available
12.1.2	BS ISO 31-4	SQUS – Part 4: Heat	Not available
12.1.2	BS ISO 31-5	SQUS – Part 5: Electricity and magnetism	Not available
12.1.2	BS ISO 31-6	SQUS – Part 6: Light and related electromagnetic radiations	Not available
12.1.2	BS ISO 31-7	SQUS – Part 7: Acoustics	Not available
12.1.2	BS ISO 31-8	SQUS – Part 8: Physical chemistry and molecular physics	Not available
12.1.2	BS ISO 31-9	SQUS – Part 9: Atomic and nuclear physics	Not available
12.1.2	BS ISO 31-10	SQUS – Part 10: Nuclear reactions and ionizing radiations	Not available
12.1.2	BS ISO 31-11	SQUS – Part 11: Mathematical signs and symbols for use in physical sciences and technology	Not available
12.1.2	BS ISO 31-12	SQUS – Part 12: Characteristic numbers	Not available
12.1.2	BS ISO 31-13	SQUS – Part 13: Solid state physics	Not available

Table A.1 Normative references (continued)

BS 8888 (sub)clause	Standard referenced	Title of the standard	Typical illustration	
7.1	BS EN ISO 128-20	TD – GPP – Part 20: Basic conventions for lines	——————	continuous line
			– – – – –	dashed line
			— – — – —	dashed spaced line
7.1	BS EN ISO 128-21	TD – GPP – Part 21: Preparation of lines by CAD systems	Not available	
7.1	BS ISO 128-22	TD – GPP – Part 22: Basic conventions and applications for leader lines and reference lines	(Ø4, 70)	
7.1	BS ISO 128-23	TD – GPP – Part 23: Lines on construction drawings	Dashed narrow line	existing contours on landscape drawings
				subdivision of plant beds/grass
				hidden outlines
7.1	BS ISO 128-24	TD – GPP – Part 24: Lines on mechanical engineering drawings	Long-dashed dotted wide line	indication of (limited) required areas of surface treatment, e.g. heat treatment
				position of cutting planes
7.1	BS ISO 128-25	TD – GPP – Part 25: Lines on shipbuilding drawings	Dashed narrow line	hidden edges
				hidden profiles

Table A.1 Normative references (continued)

BS 8888 (sub)clause	Standard referenced	Title of the standard	Typical illustration
10.1	BS ISO 128-30	TD – GPP – Part 30: Basic conventions for views	
10.1	BS ISO 128-34	TD – GPP – Part 34: Views on mechanical engineering drawings	

Illustrated index to BS 8888

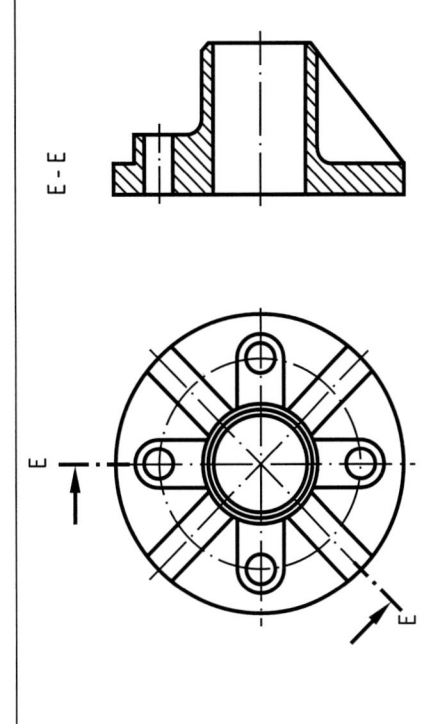

Table A.1 Normative references *(continued)*

BS 8888 (sub)clause	Standard referenced	Title of the standard	Typical illustration
11	BS ISO 128-40	*TD – GPP – Part 40: Basic conventions for cuts and sections*	
11	BS ISO 128-44	*TD – GPP – Part 44: Sections on mechanical engineering drawings*	

Table A.1 Normative references (continued)

BS 8888 (sub)clause	Standard referenced	Title of the standard	Typical illustration
11	BS ISO 128-50	TD – GPP – Part 50: Basic conventions for representing areas on cuts and sections	
4.3, 7.2, 15.2	BS ISO 129-1	TD – Indications of dimensions and tolerances – Part 1: General principles	
15.2	BS ISO 406	TD – Tolerancing of linear and angular dimensions	

NOTE Annex H of BS 8888:2008 gives preferred options for dimensioning, tolerancing and lettering.

Illustrated index to BS 8888

Table A.1 Normative references *(continued)*

BS 8888 (sub)clause	Standard referenced	Title of the standard	Typical illustration
19.2	BS EN ISO 463	Dimensional measuring equipment – Design and metrological characteristics of mechanical dial gauges	Not available
12.1.2	BS ISO 1000	Specification for SI Units and recommendations for the use of their multiples and of certain other units	Not available
4.3, 16	BS ISO 1101	TD – GT – Tolerancing of form, orientation, location and run-out – Generalities, definitions, symbols, indications on drawings	Form: — ▱ ○ ⌀ ∩ ⌒ Orientation: ∥ ⊥ ∠ ⌒ ⌒ Location: ⌖ ◎ ≡ ⌒ ⌒ Run-out: ↗ ⌁ Modifier symbols: CZ LD MD PD LE NC AC S
15.2	BS EN ISO 1119	GPS – Series of conical tapers and taper angles	*(illustration of conical taper)*
17	BS EN ISO 1302	GPS – Indication of surface texture in technical product documentation	ground $\sqrt{\text{U"X" 0,08-0,8 / Rz8max 3,3}}$

Table A.1 Normative references *(continued)*

BS 8888 (sub)clause	Standard referenced	Title of the standard	Typical illustration
15.2	BS EN ISO 1660	*TD – Dimensioning and tolerancing of profiles*	
15.2	BS 1916-1	*Limits and fits for engineering – Part 1: Limits and tolerances*	Not available
15.2	BS 1916-2	*Limits and fits for engineering – Part 2: Guide to the selection of fits in BS 1916:Part 1*	
15.2	BS 1916-3	*Limits and fits for engineering – Part 3: Recommendations for tolerances, limits and fits for large diameters*	Not available

Illustrated index to BS 8888

Table A.1 Normative references *(continued)*

BS 8888 (sub)clause	Standard referenced	Title of the standard	Typical illustration
			View / Section / Simplified
14.1	BS EN ISO 2162-1	*TPD – Springs – Part 1: Simplified representation*	
14.1	BS EN ISO 2162-2	*TPD – Springs – Part 2: Presentation of data for cylindrical, helical, compression springs*	
14.1	BS EN ISO 2162-3	*TPD – Springs – Part 3: Vocabulary*	Not available

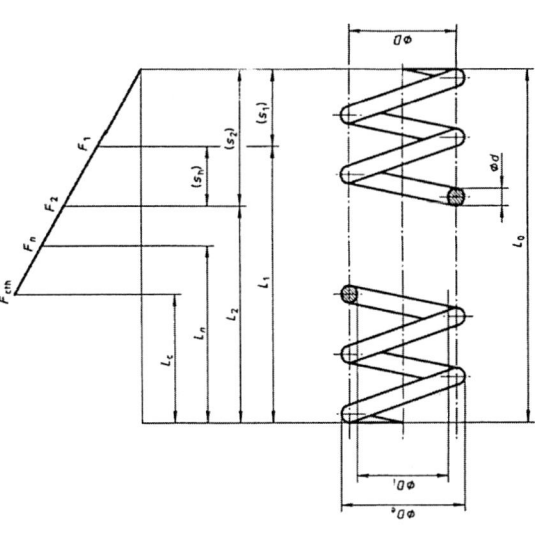

Table A.1 Normative references (continued)

BS 8888 (sub)clause	Standard referenced	Title of the standard	Typical illustration
14.1	BS EN ISO 2203	TD – Conventional representation of gears	
16	BS EN ISO 2692	GPS – GT –Maximum material requirement (MMR), least material requirement (LMR) and reciprocity requirement (RPR)	Ⓜ MMC Ⓛ LMC
19.2	BS 2795	Specification for dial test indicators (lever type) for linear measurement	Not available
14.1	BS 2917-1	Graphic symbols and circuit diagrams for fluid power systems and components – Part 1: Specification for graphic symbols	Not available
15.2	BS ISO 3040	TD –Dimensioning and tolerancing – Cones	
8.1	BS EN ISO 3098-0	TPD – Lettering – Part 0: General requirements	Not available

Table A.1 Normative references *(continued)*

BS 8888 (sub)clause	Standard referenced	Title of the standard	Typical illustration
8.1	BS EN ISO 3098-2	*TPD – Lettering – Part 2: Latin alphabet, numerals and marks*	
			NOTE BS 8888 non-preferred. Annex H of BS 8888:2008 gives preferred options for dimensioning, tolerancing and lettering.
8.1	BS EN ISO 3098-3	*TPD – Lettering – Part 3: Greek alphabet*	
8.1	BS EN ISO 3098-4	*TPD – Lettering – Part 4: Diacritical and particular marks for the Latin alphabet*	
8.1	BS EN ISO 3098-5	*TPD – Lettering – Part 5: CAD lettering of the Latin alphabet, numerals and marks*	

Table A.1 Normative references *(continued)*

BS 8888 (sub)clause	Standard referenced	Title of the standard	Typical illustration
8.1	BS EN ISO 3098-6	*TPD – Lettering – Part 6: Cyrillic alphabet*	
14.1	BS 3238-1	*Graphical symbols for components of servo-mechanisms – Part 1: Transducers and magnetic amplifiers*	
14.1	BS 3238-2	*Graphical symbols for components of servo-mechanisms – Part 2: General servo-mechanisms*	
17	BS EN ISO 3274	*GPS – Surface texture: Profile method – Nominal characteristics of contact (stylus) instruments*	*Not available*
13	BS EN ISO 4063	*Welding and allied processes – Nomenclature of processes and reference numbers*	*Not available*
17	BS EN ISO 4287	*GPS – Surface texture: Profile method – Terms, definitions and surface texture parameters*	*Not available*

Table A.1 Normative references *(continued)*

BS 8888 (sub)clause	Standard referenced	Title of the standard	Typical illustration
17	BS EN ISO 4288	GPS – Surface texture: profile method – Rules and procedures for the assessment of surface texture	Not available
15.2	BS 4500-4	ISO limits and fits – Specification for system of cone (taper) fits for cones from C=1:3 to 1:500, lengths from 6 mm to 630 mm and diameters up to 500 mm	Not available
15.2	BS 4500-5	ISO limits and fits – Specification for system of cone tolerances for cones from C=1:3 to 1:500, lengths from 6 mm to 630 mm	Not available
5.4	BS 5070-1	Engineering diagram drawing practice — Part 1: Recommendations for general principles	Not available
5.4	BS 5070-3	Engineering diagram drawing practice — Part 3: Recommendations for mechanical/fluid flow diagrams	Not available
5.4	BS 5070-4	Engineering diagram drawing practice — Part 4: Recommendations for logic diagrams	Not available
13	BS EN ISO 5261	TD – Simplified representation of bars and profile sections	Angle section L Alternate symbol: L
6	BS EN ISO 5455	TD – Scales	Not available

Table A.1 Normative references *(continued)*

BS 8888 (sub)clause	Standard referenced	Title of the standard	Typical illustration
9	BS EN ISO 5456-2	*TD – Projection methods – Part 2: Orthographic representations*	
9	BS EN ISO 5456-3	*TD – Projection methods – Part 3: Axonometric representations*	

Illustrated index to BS 8888

Table A.1 Normative references *(continued)*

BS 8888 (sub)clause	Standard referenced	Title of the standard	Typical illustration
9	BS ISO 5456-4	*TD – Projection methods – Part 4: Central projection*	
5.2.1	BS EN ISO 5457	*TPD – Sizes and layout of drawing sheets*	

Table A.1 Normative references *(continued)*

BS 8888 (sub)clause	Standard referenced	Title of the standard	Typical illustration
15.2, 16	BS EN ISO 5458	*GPS – GT – Positional tolerancing*	
16	BS ISO 5459	*TD – GT – Datums and datum-systems for geometrical tolerances*	NOTE *Annex H of BS 8888:2008 gives preferred options for dimensioning, tolerancing and lettering.*
13, 14.1	BS EN ISO 5845-1	*TD – Simplified representation of the assembly of parts with fasteners – Part 1: General principles*	
13, 14.1, 15.2	BS EN ISO 6410-1	*TD – STTP – Part 1: General conventions*	

Illustrated index to BS 8888

Table A.1 Normative references *(continued)*

BS 8888 (sub)clause	Standard referenced	Title of the standard	Typical illustration
13, **14.1**	BS EN ISO 6410-2	*TD – STTP – Part 2: Screw thread inserts*	Detailed / Conventional / Simplified — Insert
13, **14.1**	BS EN ISO 6410-3	*TD – STTP – Part 3: Simplified representation*	Hexagon head screw
13	BS EN ISO 6411	*TD – Simplified representation of centre holes*	ISO 6411 – B 2,5/8

185

Table A.1 Normative references *(continued)*

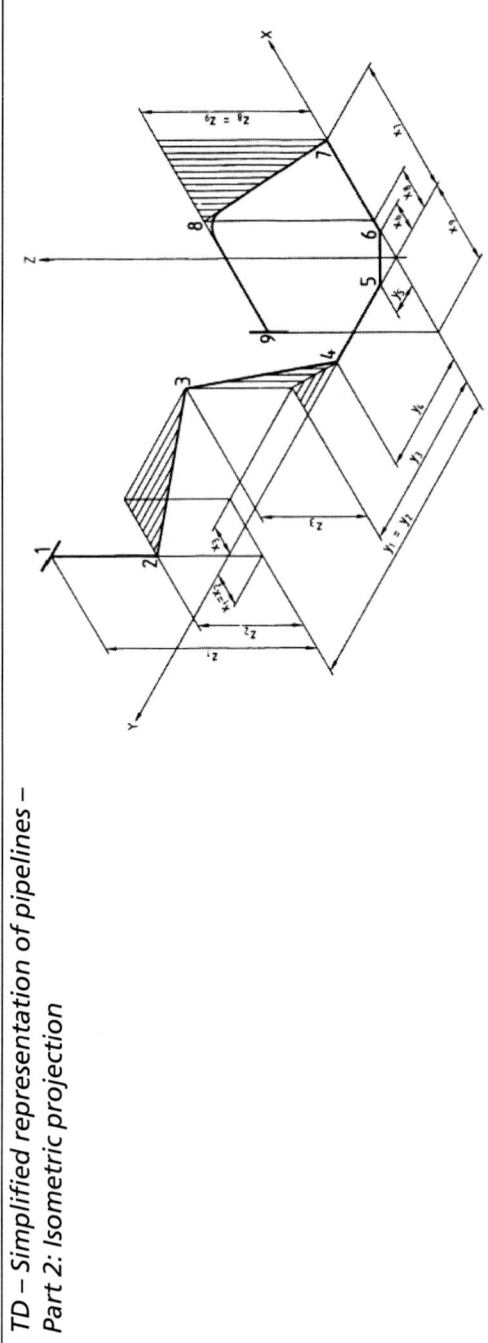

BS 8888 (sub)clause	Standard referenced	Title of the standard	Typical illustration
14.1	BS EN ISO 6412-1	*TD – Simplified representation of pipelines – Part 1: General rules and orthogonal representation*	
14.1	BS EN ISO 6412-2	*TD – Simplified representation of pipelines – Part 2: Isometric projection*	

Table A.1 Normative references *(continued)*

BS 8888 (sub)clause	Standard referenced	Title of the standard	Typical illustration
14.1	BS EN ISO 6412-3	TD – Simplified representation of pipelines – Part 3: Terminal features of ventilation and drainage systems	Scupper
13	BS EN ISO 6413	TD – Representations of splines and serrations	
21	BS EN ISO 6428	TD – Requirements for microcopying	Not available
4.1	PD 6461-1	General metrology – Part 1: Basic and general terms (VIM)	Not available
4.1	PD 6461-3	General metrology – Part 3: Guide to the expression of uncertainty in measurement (GUM)	Not available
15.2	BS 6615	Specification for dimensional tolerances for metal and metal alloy castings	Not available
15.2, 16	BS EN ISO 7083	TD – Symbols for geometrical tolerancing – Proportions and dimensions	

Table A.1 Normative references *(continued)*

BS 8888 (sub)clause	Standard referenced	Title of the standard	Typical illustration
5.2.1	BS EN ISO 7200:2004	TPD – Data fields in title blocks and document headers	Not available
15.1.1, 15.2	BS ISO 8015	TD – Fundamental tolerancing principle	Ⓔ
14.2	BS EN ISO 8062-1	GPS – Dimensional and geometrical tolerances for moulded parts – Vocabulary	Not available
14.2	BS EN ISO 8062-3	GPS – Dimensional and geometrical tolerances for moulded parts – General dimensional and geometrical tolerances and machine allowances for casting	Not available
17	BS EN ISO 8785	GPS – Surface imperfections – Terms, definitions and parameters	
14.1	BS EN ISO 8826-2	TD – Rolling bearings – Part 2: Detailed simplified representation	
14.1	BS EN ISO 9222-1	TD – Seals for dynamic application – Part 1: General simplified representation	

Illustrated index to BS 8888

Table A.1　Normative references *(continued)*

BS 8888 (sub)clause	Standard referenced	Title of the standard	Typical illustration
14.1	BS EN ISO 9222-2	TD – Seals for dynamic application – Part 2: Detailed simplified representation	
14.2, 17	BS ISO 10135	GPS – Drawing indications for moulded parts in technical product documentation (TPD)	
3	BS ISO 10209-1	TPD – Vocabulary – Part 1: Terms relating to technical drawings: general and types of drawing	Not available
3, 9	BS EN ISO 10209-2	TPD – Vocabulary – Part 2: Terms relating to projection methods	Not available
16	BS ISO 10578	TD – Tolerancing of orientation and location – Projected tolerance zone	

Table A.1 Normative references (continued)

BS 8888 (sub)clause	Standard referenced	Title of the standard	Typical illustration
4.1, 15.2	BS ISO 10579	TD – Dimensioning and tolerancing – Non-rigid parts	
20.2, 21	BS EN ISO 11442	TPD – Document management	Not available
17	BS EN ISO 11562	GPS – Surface texture: Profile method – Metrological characteristics of phase correct filters	Not available
17	BS EN ISO 12085	GPS – Surface texture: Profile method – Motif parameters	Not available
16	DD CEN/ISO TS 12180-1	GPS – Cylindricity – Part 1: Vocabulary and parameters of cylindrical form	Not available
16	DD CEN/ISO TS 12180-2	GPS – Cylindricity – Part 2: Specification operators	Not available
16	DD CEN/ISO TS 12181-1	GPS – Roundness – Part 1: Vocabulary and parameters of roundness	Not available
16	DD CEN/ISO TS 12181-2	GPS – Roundness – Part 2: Specification operators	Not available

Table A.1 Normative references *(continued)*

BS 8888 (sub)clause	Standard referenced	Title of the standard	Typical illustration
16	DD CEN/ISO TS 12780-1	GPS – Straightness – Part 1: Vocabulary and parameters of straightness	*Not available*
16	DD CEN/ISO TS 12780-2	GPS – Straightness – Part 2: Specification operators	*Not available*
16	DD CEN/ISO TS 12781-1	GPS – Flatness – Part 1: Vocabulary and parameters of flatness	*Not available*
16	DD CEN/ISO TS 12781-2	GPS – Flatness – Part 2: Specification operators	*Not available*
17	BS EN ISO 13565-1	GPS – Surface texture: Profile method – Surfaces having stratified functional properties – Part 1: Filtering and general measurement conditions	*Not available*
17	BS EN ISO 13565-2	GPS – Surface texture: Profile method – Part 2: Height characterization using the linear material ration curve	*Not available*
17	BS EN ISO 13565-3	GPS – Surface texture: Profile method – Part 3: Height characterization using the material probability curve	*Not available*
13	BS ISO 13715	TD – Edges of undefined shape – Vocabulary and indications	*Not available*
15.2	BS EN ISO 13920	Welding – General tolerances for welded constructions – Dimensions for length and angles – Shape and position	*Not available*

Table A.1 Normative references *(continued)*

BS 8888 (sub)clause	Standard referenced	Title of the standard	Typical illustration
4.1, 17, 19.1	BS EN ISO 14253-1	GPS – Inspection by measurement of workpieces and measuring equipment – Part 1: Decision rules for proving conformance or non-conformance with specifications	
4.1, 19.1	DD ENV ISO 14253-2	GPS – Inspection by measurement of workpieces and measuring equipment – Part 2: Guide to the estimation of uncertainty in GPS measurement, in calibration of measuring equipment and in product verification	Not available
19.1	DD CEN/ISO TS 14253-3	GPS – Inspection by measurement of workpieces and measuring equipment – Part 3: Guidelines for achieving agreements on measurement uncertainty statements	Not available
3, 15.1.1, 17	BS EN ISO 14660-1	GPS – Geometrical features – Part 1: General terms and definitions	Drawing → Workpiece → Extraction → Association

Table A.1 Normative references *(continued)*

BS 8888 (sub)clause	Standard referenced	Title of the standard	Typical illustration
15.1.1	BS EN ISO 14660-2	GPS – Geometrical features – Part 2: Extracted median line of a cylinder and a cone, extracted median surface, local size of an extracted feature	*Not available*
19.2	BS EN ISO 14978	GPS – General concepts and requirements for GPS measuring equipment	*Not available*
19.2	DD CEN/ISO TS 15530-3	GPS – Coordinate measuring machines (CMM): Technique for determining the uncertainty of measurement – Part 3: Use of calibrated workpieces or stands	*Not available*
13	BS EN ISO 15785	TD – Symbolic presentation and indication of adhesive, fold and pressed joints	
17	DD ISO/TS 16610-1	GPS – Filtration – Part 1: Overview and basic concepts	*Not available*
17	DD ISO/TS 16610-20	GPS – Filtration – Part 20: Linear profile filters: Basic concepts	*Not available*
17	DD ISO/TS 16610-22	GPS – Filtration – Part 22: Linear profile filters: Spline filters	*Not available*
17	DD ISO/TS 16610-29	GPS – Filtration – Part 29: Linear profile filters: Spline wavelets	*Not available*
17	DD ISO/TS 16610-40	GPS – Filtration – Part 40: Morphological profile filters: Basic concepts	*Not available*

Table A.1 Normative references (*continued*)

BS 8888 (sub)clause	Standard referenced	Title of the standard	Typical illustration
17	DD ISO/TS 16610-41	GPS – Filtration – Part 1: Morphological profile filters: Disk and horizontal line-segment filters	Not available
17	DD ISO/TS 16610-49	GPS – Filtration – Part 1: Morphological profile filters: Scale space techniques	Not available
18	BS ISO 16792	TPD – Digital product documentation – Digital product definition data practices	Not available
4.1	DD CEN/ISO TS 17450-1	GPS – Part 1: General concepts – Part 1: Model for geometrical specification and verification	Not available
4.1	DD ISO/TS 17450-2	GPS – Part 2: Operators and uncertainties	Not available
15.2, F.5	BS EN 20286-1	ISO system of limits and fits – Part 1: Bases of tolerances, deviations and fits	
15.2	BS EN 20286-2	ISO system of limits and fits – Part 2: Tables of standard tolerance grades and limit deviations for holes and shafts	

Table A.1 Normative references (continued)

BS 8888 (sub)clause	Standard referenced	Title of the standard	Typical illustration
13	BS EN 22553	Welded, brazed and soldered joints – Symbolic representation on drawings	NOTE Annex H of BS 8888:2008 gives preferred options for dimensioning, tolerancing and lettering.
15.7	BS EN 22768-1	General tolerances – Part 1: Tolerances for linear and angular dimensions without individual tolerance indications	Not available
15.7	BS EN 22768-2	General tolerances – Part 2: Tolerances for features without individual tolerance indications	Not available
19.1	DD ISO/TS 23165	GPS – Guidelines for the evaluation of coordinate measuring machine (CMM) test uncertainty	Not available
17	BS EN ISO 81714-1	Design of graphical symbols for use in the technical documentation of products – Part 1: Basic rules	Not available

This book has been produced as a companion to BS 8888, *Technical product specification – Specification*, which is the definitive standard for technical product realization. It offers straightforward guidance, together with pictorial representations, to all practitioners of technical product specification, providing engineers engaged in design specification, manufacturing and verification with the essential information required for specifying a product or component.

BS 8888 is being increasingly adopted in major sectors, particularly in defence contracts, and so the need to use the standard becomes more pressing. This publication presents up-to-date information on the technical drawing aspects of BS 8888, in a manner similar to the presentation in the old BS 308, and will help users trained according to BS 308 to adapt to the new standards.

It includes comprehensive sections extracted from and referenced to international standards relating to linear, geometric and surface texture dimensioning and tolerancing, together with the practice of welding symbology, limits and fits and thread data. It also includes an illustrated index to all standards referenced in BS 8888.

COLIN H SIMMONS IEng FIED, is an International Engineering Standards Consultant and a member of British Standards and ISO committees dealing with technical product documentation and technical product specifications. He is a former practising mechanical design engineer and Chief Standards Engineer of Lucas Industries and is the author of many publications, including the revisions of PP 8888-1, *Drawing Practice: A guide for schools and colleges to BS 8888:2006, Technical product specification (TPS)* and PP 8888-2, *Engineering Drawing Practice: A guide for further and higher education to BS 8888:2006, Technical product specification (TPS)*, co-authored with Neil Phelps.

Neil Phelps, IEng MIED, MIET is a time-served mechanical engineer currently working as a design manager within a manufacturing environment. He is a member of the BSI committees responsible for technical product realization, technical product documentation, BS 8888 and BS 8887 and is chairman of the committee covering digital product specification. He is also a member of two ISO Technical Committees responsible for technical product documentation and dimensional and geometrical product specification and verification. Together with Colin Simmons, he has revised the BSI publications PP 8888-1 and PP 8888-2.

BSI
Group Headquarters
389 Chiswick High Road
London
W4 4AL
www.bsigroup.com

The British Standards Institution is incorporated by Royal Charter
BSI order ref: BIP 2155

ISBN 978-0-580-62673-9